Fast and Flawless Systems

A decorator's guide to planning and carrying out a successful job

Pete Wilkinson
www.fastandflawless.co.uk

This book is dedicated to Charlie and Ellie. They are to me what life is all about and whatever they do in their life I will be proud to be their Dad.

Contents

Contents

Chapter 6 — Understanding paint

Paint is something we need to know about. We explore the topic of paint in this chapter.

Chapter 7 — Substrates and typical paint systems

There are a million combinations of surface, primer and preparation. Here we look at how to make sense of it all.

Chapter 8 — Conventional decorating systems

A benchmark to start from.

Chapter 9 — Masking systems

Masking has burst onto the decorating scene, here we look at masking systems and what we need to consider.

Chapter 10 — Systems for spraying a room

"What order do I spray a room?", is a question often asked. This chapter puts the question to rest.

Chapter 11 — Painting the outside of houses

This can be a real money maker if we get those systems right.

Contents

Contents

Chapter 18 — Can systems be a problem?

Now we have decided that systems are the way to go, we discuss where they may go wrong.

Chapter 19 — The last one

What is the future of decorating?

Preface

Some of you will know this is not my first book, I have written two others about decorating. The first one was called Fast and Flawless and was a book about how to get into airless spraying. I originally wrote it because there was nothing out there to give to my students at college.

I am not a professional writer, I am a decorator. My spelling is not great, I am not one to use big fancy words. I take quite a bit of time to type, I am not a professional typist either. The first book I wrote was not intended for a wider audience and I never dreamed I would sell many copies.

All these things I have found have actually turned out to be an advantage. My style of writing is easy to read (I am told) and I now get other people to proofread my books to eliminate the spelling and grammar errors.

Once I had produced the first book, I discovered two things about writing.

First, I will never retire on the money I make from the books and second, I enjoy writing them. Lucky for me, decorators seem to enjoy reading them.

I get many emails from decorators who say they had never read a book until "Fast and Flawless" came along and that they enjoyed it. Plus, they found the information useful, which is always a bonus.

With all this in mind, I wrote a second book. This one would be about pricing. Not pricing spraying but pricing any decorating work. The reason I wrote this book was that I had spoken to thousands of decorators, and they all struggled with pricing their work.

This second book was easier to write because I had more of an idea of how long the process took and I had learnt how to publish a book successfully on Amazon.

I would like to say a few words about self-publishing and Amazon themselves. Self-publishing gets a bad rap, maybe not so much these days but in the past.

Why is this?

Well, I think it is because the big publishing houses have realised that the game is up, and they will lose control of the grip they have on the industry. If you publish a book through a publisher (I have never done this, so I am not an expert) then there are quite a lot of costs, most of these will be covered by the profits from the book once it is published.

If you are a new author, then you are a big risk to a publisher, and they will be unlikely to touch you. If you are an established author but not a big seller then you will make little or nothing from your creative endeavours.

Along came Amazon and changed all of that. Now Amazon get some bad press too, sometimes. They make too much money, they don't pay tax, blah, blah, blah. I think Amazon has done an amazing thing to put the power to publish your book into everyday people's hands.

People like me.

I just want to say that this is great, and I appreciate that they have done this.

The second book was like the second album, I feared that no one would buy it, and no one would like it. I mean, a book on pricing; how dull can you get? I was encouraged by my sister, who is not a decorator, by the way. She read a proof copy and said: "I am not even a decorator or interested in pricing, but I just can't put the book down". The comment made me smile and made me realise that maybe I had another "hit" on my hands.

I published the book on Christmas Eve 2019, and it went "bestseller" within a week. To say I was happy is an understatement.

Preface

I would not have written the book without several people in my life that have influenced me and supported me. These people deserve a mention here. I will mention them in the order they came into my life.

My dad is a boatbuilder by trade, he worked on the docks and did all kinds of wood type jobs. Later, when the docks closed, he worked as a joiner for the local council, a job he hated. Not because of the actual job but because of the way the council worked. The systems at the council were different to the systems on the dock and the result was a poor working environment.

Because my dad is a tradesman, he has passed on some of his ideas. One piece of advice he gave just before I started my apprenticeship. He said that for 3 years I should do as I was told. Do not argue with the experienced men and listen to everything that they teach you, even if you think you know better. After you come out of your time, you can do the job how you see fit.

I took this advice and kept my head down for 3 years, it served me well and I learnt a lot. Still to this day, I would like to think that I am still learning, and it is one reason I have not been frightened to do new things.

Thanks, Dad.

The next person that influenced me was my old boss now sadly deceased. I worked for a company called Dennis Kehoe Ltd. It was a family firm who employed about thirty-five decorators. In his heyday, he employed over a hundred decorators.

Dennis was not a decorator himself but a quantity surveyor. He had worked at a large painting firm and learnt how to get jobs, price them and make them profitable. It was a very well-run company and I take a lot from my time there. I still refer to some things we used to do.

The contracts manager was called John Kehoe. He was only young when I was at the company, but he did an exceptionally good job of organising everything to the point that he made it look easy. We never saw much of Dennis; it was mainly John we dealt with.

Sadly, he died at a young age of cancer and I never got the chance to thank him for his fair treatment of me when I was a young apprentice.

At quite a young age I got a job teaching at a local college. My boss at the time was a chap called John Cobham. John was a decorator and a very fair and supportive head of department.

He promoted me to be in charge of the decorating department, which at the time I thought was great but, in hindsight, I feel I was too young and did not have the experience. He has had a big influence on my life, and I appreciate what he did for me.

More recently, I have met Ian and Lyndsey Crump from PaintTech decorators. I have told the story in depth in my first book "Fast and Flawless" so I will not retell it again here.

Ian and Lyndsey have changed the way I look at all aspects of decorating and business and although Ian and I have hugely different personalities, I would like to think that we have both changed for the better.

Ian is always talking about systems and how important they are. Even though I understood before I met Ian what a system was, I did not appreciate how much impact having a good system can have on your bottom line.

Thanks, guys, I appreciate it. I know I am a pain.

Another massive influence on me has been Chris Berry. I watched his "Idaho Painter" videos while I was still working at college and I used to show the students his videos to inspire them to get better at their trade.

I never in a million years thought I would go over to Idaho and teach at his academy. That was an amazing experience, and we discussed decorating and how the trade differs across the pond.

One thing that Chris is good at, and what has made B&K Painting so profitable, is his systems. He is incredibly good at breaking down a job into a profitable system and I will talk about that more in the book.

Thanks, Chris and Lisa, you have been the inspiration for this book.

Finally, one of the biggest influences on my life these days has been my wife Tracey. She does not put up with my nonsense, she always sees the bigger picture and what can be possible. I have lost count of the number of times she has said "just do it then" after I describe one of my dream ideas.

When I am feeling like "packing in" Tracey reminds me why I am doing this.

Thanks, Tracey, you are amazing.

I understand that was all a bit self-indulgent, but we all have people in our lives that shape us to become who we are. The reason I write the books and run the courses is that I would like to make a positive difference to the trade.

If I only change one decorator's life for the better, then that is a big thing. Maybe one day, you will write a book about your amazing business and I will get a mention. I hope so.

Without further ado, let us get started.

Chapter 1
Introduction

"94% of problems in business are systems driven and only 6% are people driven."

W. Edwards Deming

A whole book on "systems", that has to be boring! If this is what you are thinking, then let me say a few things.

First, I will keep the book grounded in the real world of decorating. I use the word "systems" because it's a good one. I could have called the book "the best way to paint a lounge" but it just does not quite say what I will talk about.

Second, it will not be complicated. This book is an introduction to systems, I will assume that you know nothing about it. I like to simplify things as much as I can, and I will do that here. One thing we all do is overcomplicate things because we think that by having a complicated approach, we will do better than everyone else.

I think the opposite is the case.

I show a video on one of our courses, it is an Idaho Painter video and it is called "Painting two houses in one day". Go onto YouTube now and check it out.

I will wait.

In the video, the team from B&K Painting paint two small bungalows in less than a day. The first one takes just over 3 hours. 3 hours and 8 minutes, in fact.

In fairness, they went 2 days before and power washed the house and then went the next day and did the preparation needed, the video itself is the painting process. Chris explains this in the video.

When I first saw this video, I did not believe it. "Impossible!", I thought, "it can't be done". However, I showed the video to my students at college (to inspire them) and asked them the following question.

"How come the guys in the video can do that, and we can't?"

One reason I asked the question is because I didn't know the answer myself and I thought my students might see something that I had missed.

These are the typical answers:

- It is sunny in Idaho.

- They sprayed it.
- The products and paints in America are better.
- There is no health and safety, there is a guy on the roof painting the facia in one shot.
- They are a good team.

These are all good answers. However, the two main reasons are rarely mentioned when I ask the question.

First, they have a **system**. Chris is very good at devising systems and he has it so finely tuned that each decorator has a list of what he needs to do and the timescale for it.

Second, they **believe** they can do it, they do it every single day. When I asked Chris about it, he said that "No one ever told me it could not be done, so I just did it."

I think sometimes we talk ourselves out of things and tell ourselves that it is impossible or that we cannot do it when we can. We only need to get our act together and devise a system.

There are several things that can be turned into a system. The obvious one is the preparation and paints we use. This is already called the paint system, but how many of us test our products and test our paint systems to see they do what we want them to do. I will discuss paint systems in a later chapter.

I will discuss the following subjects and put together a system for you to use. I will also encourage you to devise your own systems.

The system is everything to your business, it is going to make you more profitable, it is what will make you more productive and it is what will allow you to sell the business when the time comes.

It is easy to think that when a job does not go to plan it is the decorators who are to blame, either yourself or one of your employees or sub-contractors. When, often, it is that everyone is freestyling and doing things in their own unique and untested way.

When we start on our own, we do not think we need a system to carry out the work, we already know what we are doing, we have been doing it for years and it's all in our head.

There are two problems with that. First, you may want to try new techniques and products and experiment with better ways of doing the work. Also, when you take another decorator on you will find they have their own process in their head.

Without further ado, let's look at the world of systems for decorating.

Chapter 2
What we currently do as decorators

"I have made a career of bumbling around places, stumbling on landmarks and generally being quite haphazard and shambolic about the way I go about things."

Bill Bryson

I want you to imagine that you have started at a new decorating firm. It's Monday morning and you are on a new job. There are four of you. The job is the outside of a house, everything is being painted even the windows, which are traditional and made of wood.

You arrive at the job at 7:30 am.

Only you have arrived at the job. Mmm that's strange, I am sure that they said they started at 7:30 am. You get out of your van and have a look at the job. There is quite a lot of work to do.

But fortunately, it is a warm sunny day and the forecast is great for the coming week.

The foreman arrives at 7:40 am.

He explains that he was stuck in traffic and chats about the work that needs to be done.

"We need to prepare everything first. There is going to be two coats of masonry paint on all the render. The windows are going to be stained and the gutters and facias are being painted in oil-based undercoat and gloss. I think we should get the job done by Friday with four of us."

He does not seem worried that there is no sign of the other two. We make ourselves a brew and walk around the job to get our heads around it and discuss what we will do.

The other two arrive at 7:50 am.

They make a brew and sit in their van; they do not appear interested in the job planning process.

In fact, no one does any planning. The guys get out of the van and the foreman discusses preparing facias and windows, etc., and everyone gets on with it. By the end of the first day, no one knows if they have made good progress or not. Everyone has decided that it's a job that will take a week and everyone seems confident this will be achieved.

Each day, individuals decide what they will do. Sometimes there is a brief chat with the foreman but if everyone is busy, and the job is completed by the end of the week, he is happy.

The boss visits once to see that progress is being made, has a quick walk around the job with the foreman and is happy with the progress.

By the end of the week, the job is completed. They finish it by 1.30 pm so have "an early dart" and call it a day.

This example is a simple job, but this is fairly typical of how a job is carried out. I could have picked any job, such as redecorating a full inside of a house. Some decorators will be more organised than others and that someone working on their own will probably be more efficient than a firm.

I think we have a "make it up as we go along" approach to most jobs, myself included in this, by the way. Many of us lack some discipline as well. So, start and finish times can be flexible as can the duration of the job.

I have two questions for you to consider and these are:

1. Why do we work in this way?
2. What are the advantages of having a more organised approach?

Let's consider the first question, why do we work this way? I think we copy the methods we pick up when we start decorating. If you were an apprentice at a company, then you will automatically slot into their ways of working and then carry on without thinking much once you are a time-served decorator and you are doing your own jobs.

If you worked with a family member or a one-man band then a similar thing would happen, however you would only pick up the methods of that one person.

I don't think we ever think about the approach of the decorating jobs we do. We have decided how long everything takes and even though we may try to go a little bit quicker we would not consider, for example, cutting the week-long outside down to a day. This would be inconceivable to us.

What are the advantages of a more organised approach then? Why even bother to shake things up and try a new system or even formalise the system you already have, even though it's a little haphazard?

Well, there are two advantages, the first one is that you stand to make more money if you can make the job more efficient. In our example, I have talked about cutting a week job down to a day and this would be quite a leap to start with. However, I think with a little organisation you could easily cut the 5 days outside job down to 4 days.

This is 20% faster!

They finished just after lunch on the Friday, anyway. They all arrive late on the first day, too. So, a couple of improvements could be made.

If you cut the job down to 4 days, then that is quite a saving. What you do with that saving is up to you if you are self-employed. But you could either have a long weekend in reward for being organised or you could start a new job and make some extra money. 20% faster could mean 20% more income and that is some pay rise.

The second advantage is more subtle and would not maybe be on your radar in the early days. You may want to expand the company and have teams of decorators organised by a contracts manager. If you had a documented system for carrying out various jobs, then it would be easier for your teams to follow that and be more efficient.

If you came to sell your decorating business then the documented systems would mean you are not needed for the business to run like a well-oiled machine and that would mean that not only could you sell it but the business would also be worth more.

Chapter 3

Brush, roller, and spray

"Continuous improvement is better than delayed perfection."

Mark Twain

When you talk to decorators about improving their business and the amount of work that they could turn out, the conversation quickly comes around to brushes, rollers and sprayers. Which system is best? Does a bigger roller go quicker than a sprayer?

What is your preferred weapon of choice for emulsioning a ceiling for example? A brush? Would you prefer to roll it? Would you always spray it? Believe it or not, I have asked this question to many decorators and I have had all three answers.

What is your answer?

Are you right?

Of course, you are. You have always done ceilings this way. Would you try another approach? Probably not, and why would you?

It's a funny thing because if your answer was roller then you are thinking to yourself, "why would you brush a ceiling, that would just be so slow" or you may be thinking, "why would you bother to get your sprayer out to do a ceiling, it's just not worth it."

If your answer was brush then you will be thinking, "I have always brushed my ceilings, it gives a better finish and is more professional. Rollers are a bit DIY." Or "sprayers? really, for a ceiling? Don't be ridiculous."

Finally, the sprayer guy will be thinking, "a brush for something the size of a ceiling, really?" or "a roller would be slow, and the finish is poor, I would have the ceiling sprayed in a couple of minutes and it would be perfect."

All three decorators think they are right and all three know why any other approach is wrong. The point is not which is the best system, we need to stop getting hung up on application methods like that is the only thing that matters. The main thing is what is the best method for each circumstance.

We decide that we are a roller guy or a sprayer guy and that's the way we work, when really we should be looking at a job and thinking what methods are best to use on this job and then use a mixture of methods to do the job. You may actually brush, roller and spray all on one job because that is the best way to do it.

What is the main reason you use the method that you do? It is probably the way you have always done it and it was the way you were taught to do it by your old boss. That's it.

We have always done it this way.

A new system could be invented tomorrow that would be much more efficient and that would make us a fortune, and do you know what? I bet most decorators would not use it because that's not the way they were shown back in 1984.

Rollers

How do I know this? Because it has already happened. Rollers were invented in 1940 but never got much traction until Sherwin Williams promoted them and marketed them in the mid-1960s. This was in the States, of course.

When they first came over to the UK, they were shunned by the trade. They were seen as DIY and decorators felt that the orange peel texture was a bad finish and nowhere near like the superior brush finish. Some decorators still think this today, believe it or not.

Imagine if you could go back in time to 1970 with what you know now. You could roll every job, no one could come close to your speed and you would make a fortune.

Everyone would catch up eventually and by the eighties everyone was rolling.

In that decade, you could become a millionaire just because you were ahead of the curve.

I spoke to a decorator with a large company back then when rollers came out. He saw the potential and bought a load of rollers and buckets for his team to use. After a while he suspected that they were not using them, the jobs were not getting done any faster.

He went onto a job to see what was going on and there were all the decorators brushing the walls and ceilings with great big brushes and the rollers and buckets were sat in the corner of the job unused.

Amazing, imagine that!

He hired a consultant to go on-site and work with the team to show them how much more productive they could be with a roller. You see they were all set in their ways, they had always decorated using a brush and even though the evidence was massive that the roller was better and faster, they came up with reasons they were not using them.

"It's DIY."

"The finish is poor."

You get the idea. These were not real reasons, and we hear none of these things said now. If you said to your fellow decorators on a job that you would brush it because it was a better finish, then they would think that you were mad. Plus, you would not earn much money.

Spraying

Sometimes you can understand something better if you look at it from a different viewpoint. When working in an industry you can be a little bit close to it so you don't see the bigger picture.

Once upon a time, many years ago, cars used to be brushed to a finish. Skilled people did it, the paint was laid on quickly with brushes and the paint used to dry to an almost perfect finish. It was a real skill as you can imagine.

Now imagine if they still did that. It would be ridiculous, wouldn't it? I mean, it's so slow for a start. Imagine a modern production line with a team of hand painters at the end. Would you even buy a car that was hand brushed, even if they did try to convince you it was "high end" to have your car brushed?

You can clearly see that from a speed of production viewpoint and from a superior finish viewpoint spraying is the way forward.

Now also imagine back in the early days when they were brushing their cars what would be the reaction of the coach painters? They are highly skilled and highly paid individuals. They could get an almost flawless finish on a car. Would they change to the new way of working?

No, they wouldn't. They would harp on about overspray and the hassle of setting the sprayers up. Not many of them around these days to ask.

Quality of finish

I was chatting with someone who had recently bought a new house. Her husband had a good job, and they had paid nearly £400,000 for the house. That's nearly half a million pounds.

Now for half a mil, I would want the finish to be amazing, jaw-dropping, perfect. But it wasn't. The ceilings and walls had heavy roller marks in them the woodwork was done in oil-based paints and was yellowing. The cutting in was terrible, there was even paint on the door handles and windows. Half a mil, don't forget. This is not some £20,000 car.

She was furious with the builders, the finish was not her only gripe, but I will leave that out of the story. The builder sent a painter to put the work right. They made it worse and also got paint on the carpet.

It's pathetic really.

We discussed the work at her house and one thing that she could not understand was the roller marks on the walls. They look terrible she said. When I explained that they were the norm, she could not get her head around it.

She lost so much faith in the decorators she had redone all the work herself. Now that is bad. If you are a decorator yourself and you have worked for a builder then you will understand completely how this happens. The main point of this story is for us to realise that a house is a very expensive purchase and should have paintwork that rivals a car.

Touching up

This book is about systems and not about spraying. However, if you will put together more productive systems then spraying will form part of that. I feel that if we can produce a high level of finish on new properties then this will lift the trade as a whole and improve our image in the eyes of both the public and builders.

One of the big objections I get from both builders and decorators is that it is difficult to touch up when a wall or ceiling has been sprayed.

Often, a separate team is sent in to do the snagging, and they use brushes and rollers. By this time, the carpets have been fitted and, sometimes, the house buyer will have moved in.

But I will use the car manufacturing industry again as an example of how ridiculous the situation is. Let us imagine you have bought yourself a brand-new Ford Focus. You go to pick it up from the dealership and it has a roller finish.

It looks terrible, it's not even a good roller finish.

When you enquire about it, this is what you are told.

"Well, the thing is we are a bit rubbish at making cars at Ford, sometimes after the car has been painted the paintwork gets damaged. It's easily done, if you have a chisel in your pocket or you have to work on the car after it's painted it can get damaged. So, what we do, we touch it up with a roller and here it is."

What!?

He really said that.

Well, there are two things about the above conversation. First, if a Ford employee damaged a finished car, by accident or otherwise, then they would be in serious trouble. Their employer would not just laugh and say it can't be helped.

Second, you would not accept this as a customer. You would be down the road buying an Audi. They are spray finished and look lovely. These guys know how to make cars that's for sure.

Now a building site is harder to control than a production line in a factory, but the problem will be solved and the first builder to do it will clean up.

The main reason for this chapter is to point out how we all get stuck in our ways and if we want to improve our business and income then we need to step out of our comfort zone and change the way we work. We need to stop making poor excuses like "It's a bit DIY" and start making more money.

Brush, roller or spray? We need to use all three as part of a productive system. Stop calling yourself a "roller guy" or a "spray guy" and call yourself a successful decorator who makes good money by being organised

Chapter 4

How efficient is it possible to be?

"If you define the problem correctly, you almost have the solution."

Steve Jobs

In this chapter, I want to discuss how efficient it is possible to be. To do this, I will discuss three jobs I have been involved with and where I have seen big improvements in productivity.

First, what does it mean to be more productive? It is not really something that we talk about much. Yes, we talk about getting the job done quicker but what we usually do is just cut corners, we don't actually get more productive.

How efficient is it possible to be?

An increase in productivity is simply this. If I paint a door and it takes 10 minutes and then I use a new system and now I take 5 mins then I have become more productive. In that example, I have doubled my productivity, which is mind-blowing. It's a 100% increase in productivity.

What does that now mean to the decorator? It means that if he works 6 days a week then now, he only has to work 3 days to earn the same money. Imagine that! We would all love to work a 3-day week but here is a way to actually do it.

What else does it mean? Well, if you want to carry on working 6 days a week because you enjoy your job then you can and earn double what you normally earn.

Double! Give yourself a 100% pay rise.

Imagine that.

I will discuss some jobs I have done over the years. All three examples involve spraying, but this does not mean I think that spraying is the only way to improve productivity.

I spoke to Chris Berry about this and asked him what percentage of his work he sprayed. He said 60%. This means that 40% of his work he uses brushes and rollers. Now the answer may have surprised you, it surprised me.

This means that if you are going to get more productive then it's about more than just including spraying into your business, you will have to work better with your brushes and rollers too.

The sash window job

The following was an outside job. It was the ground floor of a big house and had twelve windows. Some windows were sash windows, and some were just standard windows. They were in reasonable condition.

There were two of us on the job so that made things easier. The weather was amazing, and we were having a nice sunny day. Now I rarely do outside work, so this job was unusual for me. I had decided that I would spray it and I decided I would try to do it in the day.

We spent a couple of hours prepping and sanding and making good. Then we masked off the glass and the brickwork. We used a 3M hand masker with brown paper and masking tape to do this. The process didn't take too long and by lunch, we had it all masked.

I used a water-based system on the job. It was an exterior water-based undercoat and gloss system. I won't mention the brand, but it was a "Weathershield" system. The system was guaranteed for 6 years. I could just as easily have used an oil-based system, I just like to experiment. Obviously, to get it done in the day I needed a quicker drying paint.

The system was one coat of undercoat and two coats of gloss. I applied the undercoat before lunch so it could be drying while we had a break.

I sprayed the first window and timed myself. It took me 30 seconds. Wow, that was great. I sprayed all twelve windows and then we had our lunch. I gave the undercoat the recommended 2 hours to dry although in the weather we were having I think it would have been dry in half an hour.

We were only working up the road from where I lived, so we nipped home and had some lunch and a break.

When we got back, the paint was dry. We abraded the undercoat to give the surface a key and also to get a nice smooth surface.

Then I sprayed the first coat of gloss. This again took about 10 minutes. If you are sharp you may be wondering how I did the edges and the like. Well, shock horror, I brushed them. I could not spray the windows open because I would have got paint in the house. I sprayed the window closed and then once it had flashed off, I opened them carefully and painted the edges.

Anyway, 2 hours later we sprayed the final coat and once it had flashed off (in about 20 mins) we demasked and cleaned up. The job looked great. We had easily done it in the day.

The customer could not quite believe it. He said that the work had exceeded all his expectations and that he had seen nothing like it before in his life.

I won't discuss the money side of things, but if you are a decorator then you will know how much you would charge to paint twelve sash windows with three coats of paint.

It is difficult to say how long it would have taken by brush, but I think it would be at least half an hour per coat per window. So, twelve windows are 6 hours. Then three coats are 18 hours.

Three days, realistically.

Okay, there were two of us, but in the afternoon there was only me working. On my own, it would have taken me a day and a half. So, I think we did it twice as fast. It was the first time I had tried this new approach, so I think I would have got much quicker. I think I could get up to three times faster than a conventional approach.

Student apartments

When I wrote "Fast and Flawless" in 2016, I talked about spraying student apartments. I will revisit them in this book because I have moved on since then.

First, I will set the scene in case you have not read my previous book.

Every year we redecorate quite a few student apartments. They are in three different towns, too. We do them in Manchester, Liverpool and Preston. We normally do about 200 apartments and these are over a few floors. They vary from site to site but typically there are seven stories.

The rooms are magnolia walls and white ceilings with the woodwork glossed. They are redone in durable matt, white on both ceilings and walls with the woodwork glossed.

The rooms have had their carpets taken out and, sometimes, the furniture taken out. These are replaced with new ones after they have been decorated.

A floor comprises three flats. Each flat has a corridor, four bedrooms and a kitchen. A floor has twelve bedrooms, three corridors and three kitchens.

I have someone filling and prepping for me and I have someone masking for me too. All I have to do is spray. When I did them 5 years ago, I could spray a whole floor in a day, one coat on both ceiling and walls. I thought this was amazing.

Had you spoken to me back then and asked if I could go any quicker I would have said no. Spraying tends to be one speed, you can use a bigger tip to go faster but there are drawbacks to that in a small room that could slow you down again. I felt I was working as fast as I could.

How long would it take to roll a whole floor of apartments? Well, I have no idea because I have never done this. That's not the point of this story.

Fast-forward 5 years, and I was back doing the apartments. In Salford this time, which was good because the first year we did them I was in Salford, so it was like coming home.

Over the years I had been getting better and better. It would be interesting to see how I did this year. To my amazement I got a whole floor sprayed in a morning.

A morning.

That means I first coated two floors in a day instead of one. What is going on here? Well, I think I had got into a system so there was little downtime between spraying. I would work my way through the floor in and out of the rooms. I had made little improvements to the way I did things.

Each improvement might shave 5 minutes off a process and all these 5 minutes added up to doing the floor twice as fast as I did it 5 years ago. The other thing I noticed is that I didn't feel like I was working any harder. It felt easier.

I would just like to point out also that the person I was working for never once pressured me to speed up. Sometimes he told me to slow down. But because I had a system I just naturally got quicker.

Luxury apartments

The next job, I had the opportunity to do some "proper" apartments. The student apartments were great, but they were just a single room with an en suite bathroom. Very small.

These apartments had two bedrooms and each bedroom had an en suite bathroom. There was a lounge and a kitchen. The kitchen had the kitchen fitted. The student apartments always had the kitchen ripped out.

All this would have to be masked. We were also taking a different approach. This time the same person was masking and preparing and painting. The time taken was more of a true measure.

Another good thing about this job was that the previous job had all been done by brush and roller, so I had some solid times to compare against.

A further difference was that, even though I would be doing some apartments myself, I would also train several decorators up to work more efficiently.

Many decorators that I trained had worked on similar jobs but using a brush and roller. I had to teach them the new process. The new process started by masking the apartment.

The problem with this I found was not the time it took to do the masking, but it was persuading the decorator I was training that he would get faster at it. They would just not have it. Some even quit at that stage. As the work progressed, I had some people that I had trained 3 or 4 months ago, and they had got better at the process.

When a new person started, I got them to speak to the "old hands" first, and they would tell them how they had got faster at the process and also how much money they were making.

This, I found, made all the difference. Instead of fighting the process they just got stuck in knowing that in a few weeks' time it was going to really pay off.

Finally, the great thing about this job is that we had a baseline time. The decorators on the previous job were completing the apartments at 1 per week on average. We knew that if we could beat this on average then we were doing well.

Even if we were 50% faster, that would mean one and a half apartments per week. That is a massive improvement. Just to put it in perspective. Let us say the profit on the job was £20,000 (I made that up completely by the way) then a 50% improvement would mean a £30,000 profit. Amazing.

By the time we were well on with the work, we found that the average was two apartments a week, three if they wanted to get stuck in and graft and sometimes four a week. While four a week was not the norm, it was still early in the process.

Remember how I improved over several years on the student apartments? Well, I think after 3 or 4 years of doing these we could have done an average of four a week.

What does decorating four apartments a week actually mean? It means £2,000 a week, that's what.

Before we leave this chapter, I want to talk briefly about belief. What we believe is possible and how it affects the way we work. I spent most of my time when I was training the decorators getting them to believe that they could turn out more work using my system. I will revisit this concept in chapter 11 under the heading of mindset.

Filling a bucket with water

This is an example of a very simple system that can shave time off a job and when multiplied up over the year can lead to massive savings.

When Chris Berry is working on an outside, he knows that he will need buckets of water. Out in the States, they use fivers. These are 5-gallon drums. Now an American gallon is smaller than a UK gallon, so a "fiver" is like a 20-litre bucket to us.

Before we go any further how long would it take you to fill a 20-litre bucket with water?

Have a think.

4 minutes? Something like that, maybe 5 mins. Can you think of a way to fill it quicker? Well, you might turn the tap full on so the water is coming out faster. I am not sure how much that would shave off the time.

In reality, you would not even think about it. You would just fill the bucket with water when you needed it and if it took you 5 minutes then that's how long it takes.

What if I told you it could be done in less than a minute? Would you believe it? How is that even possible?

The guys at B&K painting do this when they arrive at the job.

They place three five-gallon buckets on the drive near the tap. They run a hose from the tap to the first bucket and turn the water on really slow.

Then as they unload the van and set up their equipment everyone keeps an eye on the bucket. If it's nearly full when they pass it on the way to the van, they swap the hose to the next bucket. This continues until all three buckets are nearly full of water and then the tap gets switched off.

No time taken at all really, no one is stood over the bucket while it fills.

How much time has been saved, though? They do an outside every day so by saving 5 minutes a day then that's 25 mins a week.

If we multiply 25 mins by 48 weeks, then that's 20 hours. This is just one simple little process that we would not even think about. Imagine if you have 10 such processes then you have saved 200 hours, that's over a month!

The point of the story is not to tell you how to save time filling a bucket with water but to get you thinking about looking at all your processes and seeing if you can do any quicker or even eliminate them completely.

Chapter 5
The ultimate system

"We take the hamburger business more seriously than anyone else."

Ray Kroc

Anyone who knows me knows that I love McDonald's. Now it's not everyone's cup of tea and that's fine, I am not trying to sell you a hamburger. When they closed their restaurants due to coronavirus, the queue at the McDonald's near me was all the way up the road, there must have been fifty cars.

Imagine they were queueing up at your door wanting decorating. They must have got something right.

I don't just like their food, I admire how they run their business.

I am a big fan, I have read books about the business and watched films about it too.

Sad, I know.

When I wrote this book I felt I needed to include McDonald's as an example of a company that had taken a simple product, the hamburger and put a system together that turned that product into a twenty-one billion dollar business.

Just to repeat – twenty-one billion dollars!

Now we all know McDonald's these days, so we are not surprised. They are on every street corner. To be honest, they make it look easy. But imagine going back in time to 1940 and telling your mates you were going to set up a burger stand and if you got the systems right you reckon it could turnover 21 billion dollars.

What do you think they would say?

For a start, a billion dollars would be a mind-blowing amount in 1940, most people would not get their head around that. Back then that was bigger than entire economies.

This is my first main point. Even the humble hamburger is worth the effort of putting systems in place.

The founders of McDonald's were Richard and Maurice McDonald. Mac and Dick. They looked at the most efficient way to make a hamburger. They chalked out the kitchen on a car park and got people going through the process of making a burger. They watched as people moved about the make-believe kitchen and identified any bottlenecks.

This was the start of their success but not all of it.

The biggest influence on the business was a chap called Ray Kroc. Now a few things you should know about Ray. He was not a big shot businessman back then. He was a 55-year-old milkshake machine salesman.

McDonald's had ordered eight machines for use at their restaurant. Now eight machines was a big order, so Ray wanted to check these guys out. He drove over to the restaurant and parked outside nearby and just watched.

It was busy.

He then queued up and got himself a burger, it was amazing and so were the fries. He knew that these guys had a success on their hands, and he wanted to be part of it.

Ray Kroc wanted to expand the restaurant using a franchise model. Back then this was a new concept. It relied on him duplicating the system all over the USA.

They struggled at first. They struggled to get suitable franchisees, they struggled to get the fries right, but they persevered, and they eventually got successes.

All the systems were refined and documented. Each new restaurant that opened gave their staff training and got them up to speed.

There are obviously many reasons for the success of McDonald's, but their system is one of the main reasons. How do they pass these systems on to their staff?

They have Hamburger University.

This is more difficult to get into than Oxford or Cambridge and this is where the management systems and how everything is done is taught. There are several classrooms and workshops around the campus for people to learn.

Imagine knowing that your systems are so important that you set up an educational institution to teach people what to do. Also, imagine that the courses are so good that it's harder to get into than Oxford or Cambridge.

What has this got to do with decorating? Well, just as some said that it was just a hamburger business and not worth the effort, some say that decorating is not worth the effort either. Well, it is worth it and one day someone will put together a world-beating decorating company with killer systems and it will turn over £21 billion. Imagine telling your friends that in the pub.

Chapter 6

Understanding paint

"Out of clutter, find simplicity. From discord, find harmony. In the middle of difficulty lies opportunity."

Albert Einstein

Paint.

We are painters, so we know all about paint, don't we? We should do but after you have read this chapter you will realise that you do not know as much as you should.

It is a funny thing about paint and painting, it's so simple, isn't it? You speak to people and they know all about it. They have emulsioned their lounge, so they must know.

If you ask a few questions though, it does not take long before you realise it's actually more complicated than it seems.

What would you prime concrete with? It's a swimming pool so it will be underwater.

What do I prime the tiles in my bathroom with? Will the paint stand up to the high humidity? How long will it need to cure before I can use the bathroom?

What paint should I use to paint my kitchen? Should I use a two-pack paint? There is a single pack available and you can use a crosslinker, would that work okay? What would be the advantage of that system?

Is water-based paint better or solvent-based? I am based in Florida and I am painting the outside of the house.

Need I go on? There are a thousand different surfaces, a million different situations and a thousand new paints coming out every year. You do not get that with wood, do you?

If you are a decorator, how many of those questions could you answer correctly off the top of your head?

In this chapter, I will look at paint systems and the best way to understand them. Don't worry, it won't be like a textbook, but I feel that this is one of the main areas we need to get our head around so we can be confident that our systems work, and we can speak confidently to customers.

Maybe even impress them.

First, let us have a look at how paint is made. In the real world, it can be complex, but all paints follow the same basic formula. The three main components are:

- Binder
- Pigment
- Thinner

The binder, or sometimes it's called a medium, is the liquid part of the paint. If you have a simple varnish, like linseed oil, then this is the binder. If you put nothing in it then you will have a slow drying, soft but long-lasting coating. It's not very practical in the real world but it's a start.

There are many types of binder and they can be solvent-borne or water-borne. There are basically two types of binder. Oils and resins.

Oils are slow drying and flexible. Resins are hard and brittle. Depending on what you want from the coating you can use one or the other. For example, polyurethane resin is a hard-wearing resin coating. It's good for floors and areas of high wear. On a wooden window, however, it may crack because it's too brittle to cope with the wood movement.

I have discussed much of this in my previous books, so I will not go too far into it now.

A common oil-based coating is alkyd resin and a common water-based coating is acrylic resin. There has been a move in recent years towards water-based paints and there has been resistance from decorators towards them.

Traditionally, oil-based coatings are nice to use, they brush out well and they flow, which means they level out and give a good finish. They are fairly slow drying, which makes them easy to use when applying by brush.

There are disadvantages to oil-based coatings. They are slow drying so you can only get one coat on per day. They are smelly so some people don't like them in their house. You have to clean your brushes out in white spirit, which then has to be disposed of.

Finally, they give off solvents when they dry, and this goes into the atmosphere. The so-called "VOC's" that everyone talks about. Volatile organic compounds. There is a phrase to throw into the conversation at a party. The world has decided that it wants to cut down on these so paint manufacturers have had to reformulate a lot of their oil-based paints.

This has led to premature yellowing of oil-based coatings. It is especially noticeable with white paints. Sometimes, the paint yellows in months and does not look great.

Early water-based systems were not good. They were difficult to brush, and the finish was awful, this put a lot of decorators off. Modern water-based systems are getting better.

I am at an advantage because I spray my woodwork, and water-based paints spray great. They also dry quickly, so I can get a few coats on in a day.

One thing that I have realised is that most "off the shelf" paints are formulated to be brushed or rolled. They are thick and gloopy and need work to get them to spray well. If you go over to the States, you will find a whole array of water-based paints designed to be sprayed.

A paint company can design a paint that will do anything that you want. Let me illustrate this with a simple example using oil-based paints. If you are painting a bare wood front door and you are using an oil-based system you will use three paints.

You will use a primer. Then you will use an undercoat and then finally you will use a gloss. Why not just give it three coats of gloss? Well, if you have tried this then you will know that it will not work.

Primer is designed to **penetrate and seal** the wood. Nothing else. It is thin and oily and will stick to the wood and penetrate the grain. All these things are good for a primer. Because it is oily, it may take slightly longer to dry.

The undercoat is designed to **cover and build** the paint system. It has a lot of pigment in it so that the colour will be nice and solid (gloss does not cover well) and it will build the surface almost like a liquid filler to give you a good smooth surface to apply your gloss to.

Gloss is designed to **shine and flow**. It has less pigment in it so that the paint has more shine and also so it flows and levels out.

You can see that different coatings are designed to do different things. The paint manufacturer can design the paint to do anything that they want. They try to design a coating with a wide use in the industry so that they can sell a good volume at a fair price and make money.

There are paints out there that claim to do it all, they are a primer and a topcoat too. I am sure that they work but the paint manufacturer will have made compromises at each stage.

When you are going to put together a paint system to paint something that you have not painted before you are best going through three stages. This is especially important if you will be using the new system on a high value job.

1. Speak to more than one paint company. Pick three that you think will make the product you are looking for. They have a technical department whose job it is to advise you on the best paint system to use. If you speak to more than one, then you will get a better idea of what is available.

2. Once you get your three recommendations then get the data sheets for the three paints and read them. What is the solids content? What is the drying time? What is the recoat time? Is there any special preparation needed? Understand the coating.

3. Then buy the three paints and test them yourself. This is the only way to get a real feel for which you think is the best. Often, your supplier will give you a sample of coating to test if they know that you will be buying a lot of product.

Another thing to understand with paint systems is the thickness of the coating you will apply. Often, we are applying a decorative paint and the main thing is that it covers and looks good.

However, sometimes, you are applying a coating where the thickness does matter. The classic example is an intumescent paint.

When you price to decorate a job then the number of coats you quote for can become quite an issue. You need to be sure that you quote enough coats for the paint to cover and this can vary from coating to coating.

What is a coat then? How do we measure it? When you apply a coat of paint by brush, then you are looking at the wet film thickness (WFT). This can be measured using a wet film thickness comb.

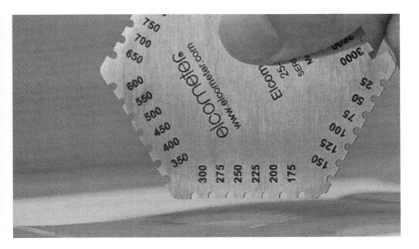

Once the paint dries it shrinks down, typically by half, but it depends on the solids content of the paint. Then you have the dry film thickness (DFT) and this can be measured also but it's more difficult. On steel, you can use an electronic device. On wood, you would have to take a sample and test it in the lab.

The paint film thickness is measured in microns. 100 microns is a tenth of a millimetre. Pretty thin. A typical coating applied by brush would be 50 microns when wet and 25 microns when dry. A coating applied by roller would be slightly thinner.

Now I know it depends on how thick you apply the paint by brush, but these are typical measurements.

If you have priced for two coats applied by brush, then you have priced for 50 microns dry film thickness if the coating has 50% solids (most decorative paints are lower).

If you apply paint by spray then it's possible to put on 100 microns (in one application) when wet and this will dry back to 50 microns, basically the same.

So, the question is, is 50 microns one coat or two? Of course, it depends on the application method but if you apply 100 microns by spray and when it dries back it has covered then that is a good job.

From a productivity viewpoint being able to apply a 100-micron coating in one go is a massive saving. I will let you go away and think about that one for a bit.

As decorators, we need to understand paint better. It is our main area of expertise, so we need to be able to talk to customers as an expert. Do not rely on what the person who works in your local paint supplier tells you. They don't know. Do your leg work and it will pay off eventually.

Chapter 7

Substrates and paint systems

"Learning new systems and processes is not mandatory...but neither is staying in business"

Bobby Darnell

One of the main things we need to know as painters is which paint to use on which surface. A "substrate" is basically a surface; steel and wood are substrates. There are many substrates out there, but I will discuss a few common ones that we paint all the time.

There are two things we need to know, the first is how to **prepare the surface** to receive the paint and the second is which **primer** to use to adhere to the surface.

Usually, if a paint system fails it is because one of those two things was not done or not done correctly.

Let's look at plastic as an example.

You need to prepare it to receive the paint and you need to paint it with something that will stick. If you just slapped a coat of emulsion onto the plastic, then it would flake or scratch off.

Plastic needs to be degreased so that there is no contamination on the surface. Then it needs to be lightly abraded so not to damage or scratch the surface. A very fine wet or dry paper or a Scotch-Brite pad would do the job. Then you need to apply something that will stick to the non-porous surface. Neat oil-based gloss works well.

Some paints are formulated to adhere and even bond to the plastic and it is worth researching the latest products.

These days, there are many "all surface" paints that claim to be able to prime any surface. Zinsser Allcoat is an example. When you look at what the paint will prime, it is like a miracle coating, and we wonder how we managed without it. I am always sceptical when I hear these claims.

If a product has been formulated to do many things then, usually, there have been compromises made along the way. If you have a paint that is a primer for a specific surface, then it is more likely to do the job well.

I want to discuss several surfaces that are commonly painted and outline the preparation and paint systems for each. This stuff is pretty basic and if you are a decorator then you should know this. However, I know you may not have a decorator background, or you may have limited experience and have not painted many types of surfaces.

Before we discuss examples of different surfaces, it is important to understand the porosity of the surface. A surface can be porous or non-porous. An example of a porous surface would be bare plaster. The surface is like a sponge and the paint soaks into it.

The advantage of this is that the paint will adhere to the surface better because it has soaked in and then once it has dried it has a grip.

Non-porous surfaces are things like tiles or plastic. The paint cannot soak in and just sits on the surface. Here, it is important that the paint is "sticky" and will adhere to the surface. Some paints have been formulated for this situation and they are called "adhesion primers". There a few of these on the market and I will discuss these in our examples.

Some substrates such as steel seem like they are non-porous, but they are porous, and the paint will soak into the surface and adhere.

So, the first thing you need to establish is the porosity of the surface or substrate. The less porous it is then the harder it will be to get paint to stick to it.

Plaster and masonry surfaces

This is probably one of the most commonly painted surfaces. It is very porous so it is easy to get the paint to stick. It is important that you thin the paint before applying the first coat.

Plaster surfaces are chemically active and if you are going to apply an oil-based system to the surface then you will need to use an alkali-resisting primer otherwise the oil in a standard oil-based primer will react with the alkali in the plaster and you will get saponification. This is a great term that you can use to impress your friends.

If you are going to apply emulsion to the plaster then you need to decide which emulsion to use. Contract emulsion is designed to paint new plaster and will allow the plaster to dry out through the coating if it is not fully dry. New build houses can take up to a year to fully dry out and for this reason, contract emulsion is recommended.

However, the drawback is that contract emulsion has no binder in it, so it is not very hard-wearing. If I were working on an extension and the plaster was fully dry, I would be tempted to use vinyl emulsion.

Whichever type of emulsion you use, it is important to thin the first coat so that the paint adheres properly. If you brush neat emulsion onto a bare plaster surface, then once dry it will just peel off in a sheet.

How much do you thin your emulsion for the first coat (or mist coat as it is known)? Different paint manufacturers vary with what they say but between 20% and 50%. I discussed this with a paint chemist at Crown Paints, and they said that if you are mist coating then you can't really overthin it. The main thing is that it soaks into the surface.

Before applying the mist coat any lumps or splashes of plaster should be scraped off. You should not use sandpaper on a new plaster wall.

Once the plaster has been mist coated then it can be filled if there are any holes or cracks.

Airless spray plaster

This is a similar surface to gypsum plaster in that it is porous but there are differences. First, it dries much quicker than traditional plaster and can be painted the next day. It can also be sanded before it is painted. It is designed to be sanded and you will not damage the surface if you abrade it. Obviously, do not use a really rough sandpaper.

Second, you do not need to mist coat airless spray plaster. If you are spraying it, then all you need is two coats to bring it to a finish. I would thin both coats by about 20% if I was spraying, depending on the product. Read the data sheet.

Exterior render and brick

Exterior surfaces that are painted are usually render or pebble dash. Although, sometimes, people paint the brickwork, I would leave the brick as it is. It will save lots of repainting in the future.

Render and pebble dash come in different depths. You can get a smooth render or a deeply textured one. If the surface has not been painted before and it is an old surface and is, therefore, a bit dusty or friable then you will need to seal it with stabilising primer. This will bond the particles together and give you a sound surface to paint.

Before you apply the stabilising primer you may need to brush off the surface with a soft brush to remove any loose material.

If it is a new surface sometimes you can get salts forming on the render (or brickwork). These are called efflorescence and should be brushed off before painting.

There are two types of paint you can use on an external render, one is oil-based and is called Pliolite. The advantage of this is that it is shower-proof immediately so that if you paint the whole side of a gable end and it rains, it will not wash all the paint off onto the drive or pavement.

The disadvantage of Pliolite is that, when you are rolling or spraying an outside, it gets on your arms, in your hair, on your face and it is a white spirit job to get rid of it.

There are several water-based masonry paints on the market and you can choose your favourite. I would thin the first coat by 20% so that it soaks into the surface much like when painting plaster. Then you can fill where necessary with an exterior filler. Once dry, apply two full coats of the masonry paint to a finish.

Be aware that paint never goes far on a masonry surface. I would tell the customer a rough estimate based on say 8 square metres a litre but keep the material price as an estimate and warn that it may take more paint than anticipated.

I know one decorator who priced an outside job, and he had priced it to take 15 gallons (75 litres, 15 x 5-litre tins) and in the end, it took double.

Exterior render jobs are ideal to spray with an airless sprayer and many decorators are scared to do this because of the dreaded overspray. In fairness, if you are new to spraying then you are more likely to get overspray because a lot of overspray is due to poor technique.

I would recommend you try it on a safe job, a detached house in the middle of nowhere, build your confidence up. I will discuss this more in the chapter dedicated to spraying outsides.

From a systems viewpoint, it makes sense to spray the majority of the render and only brush and roll any awkward little bits.

Bear in mind that if the render is heavy then you must back roll one coat to fill all the deep areas.

All the above applies the same to pebble dash.

Softwood

This is another commonly painted substrate. Softwood is porous and therefore is easy to get paint to stick to. The issues with softwood are knots and end grain.

Wood has little tubes running through it like straws (the tree uses these to draw water from the ground to the leaves) these are on the end of any piece of joinery work like skirtings or windowsills.

This is the end grain.

If these are not properly sealed with paint, then they will draw water into the whole piece of wood, and it will rot very quickly.

Again, you have two choices of system, an oil-based one and a water-based one.

Oil-based primer does penetrate the wood well and does not raise the grain the same way that acrylic primer does. However, I feel that the advantages of a water-based system outweigh this.

Oil-based primers are a separate product and are very oily and thin, it will soak into the wood well, but it can take longer to dry due to its oily nature. If you have used an oil-based primer, then really you would be better using a full oil-based system.

Water-based primers also tend to be used as an undercoat as well. This means they are thick in consistency. If you are using an acrylic primer, then you need to thin it. Look at the data sheet for advice on this. Dulux, for example, recommend 20% thinning for priming.

Once primed, the softwood can be abraded and filled and caulked if necessary. Do not sand the softwood when it has no paint on it. You can easily damage the wood surface. If there are any plaster splashes, then you can scrape these off before priming.

Once the softwood is at this stage, you can carry on with your chosen paint system.

If it is outside, then you will likely apply an undercoat and a gloss. I must admit on an outside I do prefer an oil-based gloss although there are some good long-lasting water-based glosses out there, they are just not very shiny.

If the woodwork is interior, I try to use satin or eggshell. I prefer eggshell, which is a 10% sheen. I find that most customers, however, like a satin. If using satin, then I prefer to use water-based, especially if it is white. Oil-based satins yellow so fast these days it makes them almost unusable.

I generally spray my woodwork and I use Tikkurila Helmi 30 unthinned using a 310 or 312 tip. I don't use HVLP unless the product is suitable. I will discuss this in more detail in a later chapter.

What about the knots?

Knots and knotting are a little misunderstood. In the old days, when everyone used oil-based paints, it was simple. The oil in the primer would activate the resin in the knot when you painted the softwood. To prevent this from happening you used a sealer that was not oil-based. The favourite is shellac knotting.

Knotting is a methylated-spirit-based coating that also seals the knot with the shellac. You should apply two coats and allow an hour to dry in between. Another method was to use a spirit-based aluminium primer. These days, people use Zinsser Bin, which is a pigmented shellac primer. This works well but is a very expensive approach.

When you read the data sheet for shellac knotting, it says that it is usually used with oil paints. This makes sense because a water-based coating will not activate the resin in the knot.

However, you can get discoloration of the white painted surface from the knots and knotting should prevent this.

Another problem with knots, especially on an outside where you get warm sunshine on the surface, is that the resin can come out of the knot and damage the paint surface. Shellac knotting will not stop this and the only real way to solve the problem is to drill the knots out and plug them with wood. Don't worry a joiner would normally do this.

Thankfully, softwood windows are a rare thing these days and they are either hardwood or uPVC.

Hardwood

Hardwood includes woods like oak, teak and mahogany among others. The advantage of hardwood as a building material is that it is both very durable and also nice to look at. The disadvantage these days is that it is very expensive, so other products such as plastic (made to look like hardwood) get used.

Because it is so aesthetically pleasing to look at you would wonder why you would ever paint hardwood. In fairness, a lot of hardwood is varnished or stained so that you can see the natural beauty of the wood.

There are several reasons hardwood is painted, firstly there are instances where hardwood needs some additional protection. Wooden boats are made from hardwood and the hull is painted to give added protection and also to make the surface difficult for weed and barnacles to stick to.

Another reason is fashion. At the moment, it's not very fashionable to have hardwood windows, people are painting them dark grey so that they look modern.

The difficulty in painting hardwood is that it is close-grained so it's more difficult to get the paint to penetrate and adhere and also many woods (teak for example) contain natural oils that help protect the wood but also mean it's more difficult to get the paint to stick.

A final difficulty is that oak contains tannin, which will bleed through a water-based system and needs to be stain blocked.

When painting hardwood you need to first degrease it with white spirit, this will help remove some of the surface oils. Once dry, you can gently abrade the surface. Because hardwood is harder than softwood, you can do this without damaging the surface. Obviously, sand "with the grain" and not against it.

I had one apprentice who I taught at college that while on a job with his employer sanded several hardwood fire doors across the grain and destroyed the lot. They all had to be replaced, a very expensive lesson for the poor apprentice. Although, I have to say, I think I would have noticed that I was damaging the door.

The best primers for hardwood are oil-based ones. Aluminium primer is the best. This is a high adhesion primer that also has stain blocking properties. Once dry, you could continue with a water-based system. I would use an adhesion primer again over the aluminium primer and then finish with whatever system I wanted to use.

If it was inside and the finish was satin, I would use a Tikkurila Helmi 30. If I was painting the hardwood outside, I would go for an oil-based system using undercoat and gloss.

Metal — ferrous

Ferrous metal is basically iron and steel. These days mainly steel. Ferrous just means that the metal will rust. Rust is iron oxide created on the surface of bare metal when exposed to the air and moisture.

We rarely paint new steel. It has usually already been painted, and we are just repainting it. If you are repainting the steel it can just be treated like any other painted surface.

If there are patches of rust on the painted steel, then these need to be scraped off and sanded back to clean steel. These areas then need to be primed with a rust-inhibiting primer such as red oxide or zinc phosphate.

I have a steel barge and I built it from new, so I have quite a bit of experience painting new steel. When sheets of steel come out of the foundry it has a layer on it called mill scale. This forms when the steel cools and needs to be removed before the steel is painted.

If you paint over the mill scale, then it will flake off as the mill scale comes away from the steel surface.

There are two ways to treat mill scale. The best way is to sandblast the steel back to shiny metal and then apply the primer. The best primer for new steel is zinc phosphate.

Another way is to allow the steel to weather, the mill scale will be shed, and you will have a thin layer of rust on the surface. This rust can then be removed by mechanical methods (a wire brush on a drill for example). Once the steel is clean, then you can prime it.

Steel is a porous surface so that the primer takes well. I would use an oil-based system on steel simply because water-based systems can cause rust. Having said that, I once painted a steel bridge with a specialist water-based system.

The bridge was sandblasted, and then we primed it with a water-based primer. The primer went on pink and then dried a deep red. It was a Belzona product and was very expensive. I could not quite get my head around why the water in the paint did not rust the newly blasted steel, but it worked a treat and the bridge looked amazing when done.

The finish paint was a water-based metallic paint that looked really good. It was meant to be "self-healing" which meant if the paint got slightly damaged then it would knit back together. I was not overly convinced about that but that's what the paint company claimed.

Metal — non-ferrous

Non-ferrous metals are all other metals that are not iron or steel. A few examples are copper, lead, brass, gold and silver. Copper is widely used in the building industry although less so these days with the use of plastic pipes and push-fit systems.

The main problem with copper (and other non-ferrous metals) is that they are not as porous as steel and will need to be degreased before painting. Also, you need to abrade the surface with some emery paper to give the surface a key.

Although these metals do not "rust", they still will corrode. Copper, for example, will form copper oxide, which is a green looking substance that forms on the unprotected copper.

The best primer for copper is zinc chromate. This is an oil-based coating that will need a good 24 hours to dry before finishing the surface with your chosen paint system.

Galvanised metal

Galvanised steel is common. The galvanised coating is zinc that is put onto the surface of the steel to protect it. It is a sacrificial coating, which means that it will corrode before the steel does. Sometimes, it is not painted and just left as an industrial looking finish, which is all the rage these days.

If you are going to paint galvanised steel, then you have two options to prepare it. You can leave it to weather for six months if it outside. This will roughen the surface slightly so that the primer will key to it.

Alternatively, you can degrease it with white spirit and then give it a coat of etch primer like T-Wash. This is a mild acid that bites into the surface to give the primer something to hold on to.

I would prime the surface with zinc chromate primer. You can then finish the surface in your chosen paint system. There are several water-based systems that you can use on galvanised steel if you like. Crown Trade water-based metal primer is one example of many.

Tiles and glass

These are two very similar surfaces and they are completely non-porous. Because of that, it is very difficult to get paint to stick to them. You cannot abrade them because they are too hard so you must make sure that they are completely clean. I used to do work that involved gilding on glass, this is applying lettering to glass and then filling in the outline with gold leaf.

When doing this, the glass had to be completely clean. I used methylated spirit and chalk powder (or whiting as it is called). This made sure that the glass was completely grease-free. Then once the lettering was complete, I painted carefully around the edge with gloss varnish. Gloss varnish is very sticky and will adhere very well to the glass.

Oil-based gloss is another paint that will stick very well to glass and tiles. These days you can get specialist adhesion primers for tiles and glass, for example, Johnstone's Speciality Ceramic Tile Primer.

Melamine

Another non-porous surface very much like glass. Again, degrease the surface, do not abrade, and use a specialist melamine primer. You can use an adhesion primer and my favourite is Whitson's Adhesion Primer, it sticks really well and sprays really well if you are a sprayer. It is water-based too, which is a bonus.

Varnish

Okay, I know this is not a substrate, but I felt it needed to be mentioned.

Earlier we talked about hardwood and said that often these would be varnished or stained. If you are going to varnish a new surface then you need to thin the varnish down quite a lot for the first coat so that it penetrates, say 50%, then the second coat thin 20%, and then finish with two or three full coats to get a nice build of varnish and an amazing finish.

Using sandpaper and abrading the surface

"Pete, chuck that yellow stuff in the bin, what are you doing?" This was a comment said to me by the foreman on a job a few years ago when I was sanding down some external windows for repainting.

Many painters skip or skimp on the sanding stage. It's time-consuming, and time is money. Skip the sanding and you make more money, the thinking goes.

When it comes to adhesion, abrading the surface is the number one thing to do. Never mind your fancy paints that claim to stick without sanding, do not skip this stage.

When I was at college and I had time on my hands, I did a little experiment in paint adhesion. First, I glossed a panel with an oil-based gloss and let it dry overnight.

Then the next day I sanded half the panel well and left the other half unsanded. Then I painted over both the sanded and unsanded areas with Macpherson's acrylic primer undercoat. Not an adhesion primer, I have to point out.

Then I let the undercoat dry.

I did an adhesion test with a key and found that the unsanded half just scratched off easily, no adhesion at all. The sanded half surprisingly was stuck firm.

I was at a training day at Crown Paints and the chap leading the day was also the troubleshooter at Crown. If a paint contractor rang up with a problem or compliant, he went out and investigated. In many cases of paint peeling, the cause was insufficient sanding of the surface.

Not all surfaces are abraded, as previously discussed. However, if you are going to abrade the surface then do it right and don't cut corners.

Chapter 8

Conventional decorating systems

"Everything must be made as simple as possible. But not simpler."

Albert Einstein

If you are a decorator, then you will know what order you carry out the painting of the various surfaces in a room so you could probably skip this chapter. However, the reason I am going to discuss this is two-fold.

First, it is a basic system that you already use so it is a good starting point for discussing systems that you are less familiar with.

Second, you may do the painting in a certain order but do not know why you do, so I will discuss the reasons for the traditional system. The reason for this is that when we change the way we work we must understand what will work and what won't.

We are going to decorate a lounge, it has been previously painted, so it is a redecoration. The specification is as follows:

1. Ceiling, two coats of emulsion.
2. Walls, two coats of emulsion.
3. Feature wall, wallpaper.
4. Woodwork, two coats of satin (oil-based).

What order would we do this?

1. Put a dust sheet down in the hallway from the front door to the lounge.

2. Remove everything that you can from the lounge. Sheet up anything that you cannot remove.

3. Sheet up the carpet.

4. Tape around the carpet to the edge of the skirting.

5. Prepare everything, sand down the woodwork, fill the walls and ceiling. Caulk if needed. Then sand down the filler once dry.

6. Clean out all the dust (even if you have used a dustless sander there will be some) and hoover up if needed.

7. I would give the sheets a shake outside and re-sheet at this point too.

8. Apply lining paper to the wall that is going to be wallpapered. This is hung horizontally.

9. Emulsion the ceiling with the first coat. Cut around the edge of the ceiling, go onto the wall slightly when you do this with a brush and then roll.

10. Cut in the wall areas, you do not have to cut in neat to the ceiling with the first coat and you can go onto the woodwork (slightly). Then fill in the wall colour with the roller.

11. Apply the first coat of satin to all the woodwork. When painting the panel door, paint the panels first and then the rest of the door. (There is a proper sequence for this, it allows you to keep your wet edge going).

12. Everything has now had a first coat.

13. Finish the ceiling and allow it to dry.

14. Finish the walls in emulsion, cut in carefully to the ceiling to get a nice sharp line. Allow to dry.

15. Finish the woodwork with your last coat of satin. Cut in the top of the skirting to the wall to get a nice sharp line. Allow to dry overnight, oil-based paint takes longer to dry.

16. Wallpaper the feature wall last. Trim to the finished ceiling and woodwork, make sure that you wipe off any paste on these surfaces otherwise the paste will strip off the paint over time.

17. Check for any "snags" and make them good. Then tidy up the job and replace any items removed.

18. Get paid.

Wow, that was quite a long system. It seemed so simple when you are doing it.

Why do we emulsion the ceiling first? Well, there are two reasons. First, you do not want to splash your finished wall with the ceiling paint, splashes tend to fly downwards.

Second, it is easier to cut in along the top of the wall to the ceiling than along the ceiling to the top of the wall. It's almost impossible to cut in upside down without using tape.

The same with the skirtings, it's easier to cut along the top of the skirting against the wall than to finish the skirting and cut the wall into them using a brush.

This is a simple system for decorating a room. If you were doing a new build, and the walls and woodwork were bare wood, then the only real difference is that you would mist coat the ceilings and walls and prime the woodwork before doing any caulking, filling and sanding.

To sum up, any system needs to be written down. It is best to write down systems that you already do so you can look at them and improve them. In time, anyone could follow your documented systems.

Chapter 9
Masking systems

"Great things are done by a series of small things brought together."

Vincent Van Gogh

"You are not a proper decorator if you mask, can you not cut in?" I am told this so often. At first, I used to argue my case: I am a proper decorator, I am time-served, I have my Advanced Craft in Painting and Decorating (and I got a distinction), I have been decorating for nearly 40 years, I have taught decorating.

After a while I started saying "Yeah, you're probably right."

The question is then — Is it DIY to use masking tape? The problem with this question is that people who do their own decorating use tape because they cannot cut in a nice clean line, they just do not have the skill. Because of this, some professional decorators frown on any attempt to use masking tape.

Another factor is that masking tape has come a long way over the years. Gone are the days when the only tape available was the cream crêpe masking tape that is a nightmare to use. These days there are a whole array of tapes on the market for every situation.

I think we sometimes forget as well that we are not the only trade that uses masking tape. The car industry uses a lot of it when they are respraying cars. I don't imagine you hear a car sprayer say that using tape is DIY.

I think that tape can revolutionise the trade for us. It can make our lives easier, it can help us make more money and it does a great job. I do not care how good you are at cutting in, if you have a white skirting and a dark blue wall then the laser line that tape gives you is amazing.

I am all for using tape.

As a decorator, I spray a lot of what I do and for that reason, I use a lot of tape. The funny thing is, though, now that I am used to using tape and I know what I am doing with it, I would use tape even if I was using a brush or roller.

Here are a few examples where I would use tape even when rolling.

1. Kitchens. I would wrap the kitchen using "tape and drape" so that it is protected from splashes. Rollers are actually messier than a sprayer. You get big blobs and heavy splashes from a roller. If the customer sees that you have protected their kitchen then they will not even be looking for splashes when you're gone.

2. The bath is the same. Tape and drape is ideal.

3. Light switches. I always tape these, it is much quicker to paint around them, and you get a neater job.

4. Light pendants. The same as light switches.

5. The carpet. I always tape the carpet off when painting the skirtings. It means you do not have to worry about getting paint on the carpet and it also means that you are not picking up bits of fluff onto your brush and putting it on the skirting.

What about windows? Well, in fairness, if I were going to brush the window (I rarely do) then I would cut it in. Even though I think it is faster to do it with tape. It depends on the number of coats. If it's just undercoat and gloss then I would cut in.

So, let's talk about tape. Tapes come in different sizes, but the common ones are 25 mm, 38 mm and 50 mm. In old money, these are 1", 2" and 3", much simpler. You can obviously get wider tape and you can get narrower tape, it all depends on what you are doing.

For this discussion, I will compare prices of 25 mm tape so that we are comparing like with like. 50 mm tape would be double the cost. I will just discuss rough prices, too. If you bulk buy, then you will get your tape cheaper. If you shop around, you may get a good offer.

Here are the various types of tape out there.

Standard tape

This is your bog-standard cheap-as-chips tape that you can buy anywhere. It is what most people think of when you discuss masking tape. The big advantage of this tape is that it is cheap, about seventy pence a roll.

The disadvantages are that it will dry onto the surface, most standard tapes are only meant to stay on the surface for about 4 hours. You cannot get a sharp line with standard tape either, the paint will cause the crêpe paper to crinkle and the paint will creep. The tape is hard to pull off the roll too, especially if it is old tape in which case it will be impossible.

This tape is good for bread and butter work like light switches and carpet. I would probably use this tape 70% of the time.

Precision tape

This is a more sophisticated tape and you pay for that. There are many brands of precision tape out there these days and you will pay about £4 a roll for it. So, it is very expensive. It has a few advantages, though.

It pulls off the roll lovely, it does not dry onto the surface either. With some precision tapes, you can leave them on for up to six months. The big advantage is that you get a sharp line when using it, so I would use it if I were getting a line between the skirting and the wall, especially if the wall was a dark colour.

Low tack tape

I am not a big fan of this. The theory is that if you are taping a delicate surface such as wallpaper you should use a low tack tape otherwise the tape could damage the paper. I think if I felt that tape would damage the paper then I would use another method. I find that the tack can be a little too low and the tape falls off. Try it yourself and see what you think.

Just a quick note on colours, the low tack tesa tape is pink, however different brands use different colours so it can get confusing. Settle on a brand yourself and then stick with it. My favourite is tesa ("tazer" not "tessa") as you can see, but experiment yourself, don't just take my word for it.

Tape and drape

This is polythene with tape on the edge. It comes in various lengths and opens out to that size. My most used size is 1.8 metres, but the 2.2 metres one can be useful too. If you want it shorter you just do not need to open it out as much.

This is very useful, and I wish I knew about it 20 years ago.

Duct tape

If you are taping anything difficult like brick or stone, then duct tape is great. It sticks well, and it is widely available. You need to be careful that you do not damage the surface because it is very sticky. Try it first and a test piece if you are worried. Things like brick will be fine, though.

This is not an exhaustive list of tapes, but I would not get too bogged down with too many types of tapes and different brands. I basically use two types of tapes most of the time, standard and tesa precision.

Hand masker

A hand masker is a tool that puts paper and tape together. It can also put plastic and tape together. You can use any combination of paper and tape to suit your needs. It is also much quicker to mask something off using this tool.

This is essential kit and there are a few on the market. My favourite is the 3M hand masker and they are about £50. You need the ladder hook that comes with it, this is essential.

Above — The ladder hook is essential

Masking systems

Above — The Trimaco hand masker

Using the hand masker is a knack, buy one and experiment with it. You will quickly get good with it and then wonder how you ever managed without it.

The ladder hook that comes with the 3M hand masker is essential to be able to hook it to your overalls while your hands are full with the masking paper and you are actually doing some masking.

Masking systems are part of our job these days, so you need to understand them and experiment a little to get your own system. Most suppliers will give you a roll of new tape to try out so there is no excuse for trying new tapes.

Chapter 10

Systems for spraying a room

"Improvement usually means doing something that we have never done before."

Shigeo Shingo

In chapter 8, we looked at a straightforward system of decorating a room. It is what we were taught at our old company and it is what we all do.

One goal of this book is to get more productive and one way to do this in the world of decorating is to spray.

Now, I have been involved with spraying long enough to know what all the objections are, and, in this chapter, we will look at some of them. One thing that holds decorators back from having a go at spraying is not knowing what order to paint things in.

I am repeatedly asked how I would spray a room if the walls were a different colour to the ceiling. Sometimes, I am told it's impossible and, sometimes, I am told that it is slower to spray.

I have written a whole book on spraying, so I won't cover the same old ground again but, just in case this is the first book of mine that you have read, I will discuss different spraying systems.

Conventional

This is a compressor and spray gun. Imagine someone spraying a car and that is the type of system.

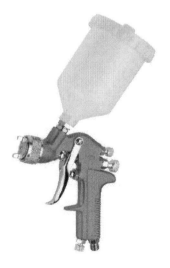

Above — a conventional spray gun.

91

The advantage of this system is that it is cheap to get going. For £150 you can get a compressor and spray gun and start spraying things. Not great for large areas such as walls but could be used for uPVC or kitchens.

HVLP (High volume low pressure) turbine

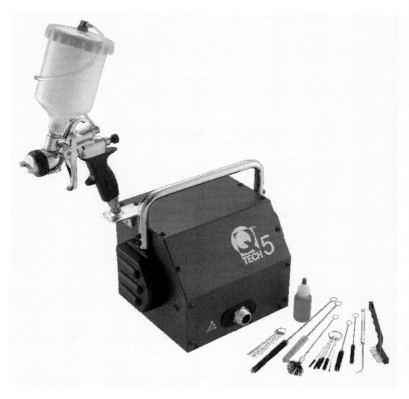

This is ideal for spraying smaller things like doors and windows. It is only good for thinner products like Zinsser Bin or Kolorbond Original and it will struggle to atomise thicker paints designed to be brushed and rolled.

The advantage is that it is very portable. However, a decent system will cost you £1000.

Airless

This is a high-pressure system and can spray large areas such as walls and ceilings. Ideal for massive spaces like a warehouse or large areas of cladding. With experience, it can also spray smaller areas such as doors and windows. This system will spray any kind of paint easily without thinning.

A good system will cost you between £1000 and £2000.

Search "Pete Wilkinson Decorators" on YouTube and you will see me spraying some stuff.

Air-assisted airless

Finally, you have air-assisted airless. This is designed for spray booths. However, some decorators do use it on-site. The advantage of it is that you get a finer finish, so it is good for spraying kitchens.

It is an expensive system and difficult to get your head around. I would only get one of these if I had a workshop and I was specialising in spraying furniture and kitchens.

The next question commonly asked is, "can you spray a domestic type job or is it just for large commercial work?"

Traditionally, spraying was done on big stuff, like a warehouse. The increase in productivity on this kind of job is immense, easily four times faster if not more.

Also, the skill level needed to spray that kind of work is low. You are not worried about overspray on the carpet on that kind of job. No real need to have good technique either, overspray in a big empty warehouse is not a problem.

Once you know what you are doing, it is actually easy to start spraying on domestic stuff. I more or less always spray my ceilings now, no matter how small. Masking takes seconds and the ceiling looks great once sprayed. It takes a couple of minutes to spray a ceiling.

Once people are happy with this answer then the next big question is, "what order do I do a lounge if I am spraying it?"

This is where your systems kick in. I have a few ways of doing this and I will look at each method in turn. I will explain the pros and cons of each method. I will assume that the room is bare walls and ceiling and bare wood.

Spraying woodwork can be up to ten times faster than a brush and the finish is amazing. I almost always spray my woodwork these days.

The first method I am going to call "My preferred system" because that is what it is.

All my systems use an airless sprayer by the way.

My preferred system

I would mist coat the ceiling and wall first with thinned down emulsion. Then I would prime the woodwork with acrylic primer.

Go home and let it dry.

Then fill, caulk and prepare all the surfaces. Make sure everything is sanded smooth and all defects are filled, and the caulking is spot on. It is unlikely I will want to do any filling at a later stage unless it is something I have missed.

Then first coat the walls with the colour. Try not to get it all over the skirting. Once you get good, you can stop an inch short of both the top of the skirting and below the ceiling.

Spray the ceiling to a finish. Two coats if needed.

Spray the woodwork to a finish. If it's a water-based satin (it usually is) then I would give it two coats with an hour in between to let it dry off, assuming reasonable drying conditions.

Leave it overnight to properly dry.

Finally, I would mask the top of the skirtings with a decent precision tape and then brush and roll the last coat.

What?! Brush and roll?

Yes, I bet that surprised you didn't it? I find there are several advantages to rolling the last coat. It saves a shed load of masking and it makes it easy to touch up the walls if needed.

However, if your customer wants a spray finish on the walls (some do) I will look at this in the other systems.

If you get a decent roller and you are a good decorator, then a roller will do a good job with emulsion.

Reverse system

This is doing the reverse of a conventional system and it allows a spray finish on the walls.

Mist coat the walls and ceiling and primer the woodwork and prepare, the same as my preferred system. From now on, I will assume that we are at this stage as all the systems start this way.

Spray the woodwork to a finish and allow it to fully dry, preferably overnight.

Mask off the woodwork with precision tape and paper using a hand masker.

Then spray the walls to a finish. Two coats if necessary. Allow to dry.

Then mask the walls off using tape and drape. Go around the top with the tape and drop the plastic right down to the skirtings. This actually does not take as long as you think, and it is what I would do if I was spraying a bathroom ceiling and the walls were tiled.

Spray the ceiling to a finish, two coats if necessary.

Then demask everything.

Above — the walls masked with tape and drape

The advantage of this system is that everything is sprayed. However, you have spent a lot of time masking and used a lot of masking materials. Plus, there is a danger of pulling the emulsion off around the top of the wall where the tape was stuck and also pulling the satin from the skirtings.

The skirtings are less likely to pull if you have prepared them correctly and the satin is fully dry.

I never use this system, but it is out there, and some decorators use it.

Shield system

If the customer wants the walls to be a sprayed finish and the walls are also a different colour to the ceiling, then this is how I do it.

It is tricky and it takes a bit to get good at it but it's worth the practice.

First coat the walls, and then finish the ceiling and finish the woodwork. Allow all this to dry preferably overnight.

Then mask off the woodwork with paper as we did in the previous system. For the next stage, you will need a spray shield.

Kraft spray shield holder

Using a roller pole for reach

There are several spray shields on the market and everyone prefers different types. I find that the metal ones are a little heavy and get loaded up with paint really quickly. Plastic ones are lighter but, because they are not absorbent, they too get dripping with paint within minutes.

Cardboard is the best. You can buy cardboard shields in packs of ten. I know this seems daft when cardboard is plentiful, but they are all the same size and you get used to how big they are. They usually come with a holder and pole too.

You can just use a piece of card and hold it with your hands, too. This will work. However, I prefer to use a holder like the one in the picture.

This is a Kraft Shield Holder, available on Amazon. The advantage of this is that it is adjustable and can be put onto a normal roller pole for reach. I just use the small Purdy roller pole.

You will need to cut in around the ceiling line with a brush, this does not have to be wide because all you are doing is painting to cover the thickness of the cardboard shield.

If you used the shield without painting a line, then you would get a thin line where the paint has missed due to the thickness of the cardboard.

Then put the shield against the ceiling and spray side to side, angle the gun down slightly to that you do not throw loads of paint onto the shield. If you are good at this, then you will get hardly any paint on the cardboard shield.

This takes some getting used to, you are spraying with one hand and you have the shield in the other hand. Once you get experienced, you can actually slide the shield along and spray at the same time.

If I am using this system, I spray side to side instead of the usual up and down. I find that unless you are good with the gun it is easy to throw paint up onto the ceiling. This is an exercise in gun control, and you will find that you get better and better at it. The angle of the gun needs to point down slightly so that none goes on the ceiling.

Once the walls are finished, I would demask the woodwork while the paper was still wet so that it is easy to demask. You can leave it to dry but you increase the chance of the tape pulling the woodwork.

Using a shield when spraying is a handy skill to have and the better you get with it the more you will use it. It will save you loads of masking once you are a wiz with the shield.

Wrap the woodwork system (one colour emulsion)

If you are spraying the walls and ceiling the same colour, then you can finish the woodwork first and then wrap it with paper. Spray the ceilings and walls and then demask. Easy.

Mask the ceiling system (upside down masking)

A final system I have used a couple of times because it is very similar to the traditional approach.

First coat the walls, then coat the woodwork and then finish the ceiling. Allow to dry properly. If it's warm in the room then it would be dry in a couple of hours.

Then mask the ceiling with the hand masker using paper.

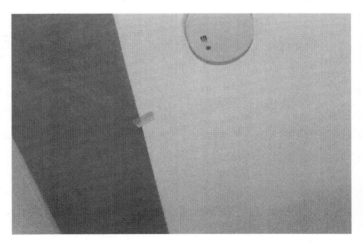

Ceiling masked with paper

This is hard because you are basically masking upside down and the paper wants to fall on your head all the time that you are working. But, as always, you get good at it.

Once the ceiling is masked, then you can spray the walls. Carefully control the gun so you do not throw too much up onto the ceiling or the overspray will go past the paper and onto the ceiling.

I would spray side to side near the ceiling and then just finish off up and down but keeping the gun angle square and not fanning. You should not fan anyway so this is great practice.

Let the walls dry overnight. Then with 100 mm tape mask off the walls above the skirtings and around the architraves and spray the woodwork.

I do not use this method much because I find that the emulsion pulls with tape easier than it does on the satin. An alternative, especially if the woodwork will be done in gloss is to spray the undercoat but brush the final coat of oil-based gloss. This way, no masking is needed.

I hope that I have not confused you with this last one. I use my preferred system 90% of the time. However, it is great to experiment. Every system has advantages and disadvantages.

As you get better with the spray gun, you may change systems. The shield method, for example, takes much more skill than my preferred method. Maybe practice with a shield on your own house.

Suspended ceiling

This is not a system but it's worth mentioning here because it is such a common situation. When the room has a suspended ceiling and only the walls are being painted then you have to mask off the ceiling.

The easiest way to do this is to slide back each tile along the wall edge or even remove the row of tiles next to the wall.

Then mask off the track with 2-inch (50 mm) tape. You can use 1-inch tape or even inch-and-half, but slightly wider is so much easier.

Suspended ceilings are actually a doddle to mask this way.

Some final thoughts on spray systems

This chapter has covered the most common question that I am asked about spraying. Every situation is different and, depending on the kind of work you do, you may find your own system of doing things.

You need to weigh up the speed, finish and masking balance. Sometimes you will spray 20% of the time and brush and roller the rest. Sometimes you will spray 80% of the time and brush and roll very little.

I advise people new to spraying to phase in the amount they spray so that they get confident with it, and their skill improves. The problem, sometimes, is that spraying is so much fun that you will want to spray things beyond your ability. While it is great that spraying is fun, try to keep your head and do things in a way that is actually better.

As your skill level improves over time you will find that you can spray most things quicker and easier than with a brush and roller. You will also learn to recognise those situations where brushing is easier. However, I always do a little time analysis to be sure that I am right, especially if it's a big job.

Chapter 11

Painting the outsides of houses

"Perfection is not attainable. But if we chase perfection, we can catch excellence."

Vince Lombardi

Chris Berry (The Idaho Painter) has been painting outsides for 18 years and it is the main plank of his business. While he will do inside work in winter, the outside work is where he makes his real money.

How much money?

Well, a team of four painters will typically complete five outsides a week. That is an average of 1 per day. Obviously, there is overlap, and a couple may get prepared at the end of a day when an outside has been completed.

Typically, Chris will charge $5,000 for an outside depending on how big it is. A smaller one would be less and a bigger one more, but this is the average. This means while doing outside work in summer, he is turning over $25,000 a week. No mean feat.

What has all this got to do with you? Well, outsides can be a good source of revenue for the small decorating business.

What is his secret?

His secret is that he has a system for doing the outsides and all the guys that work on the job know the system.

Oh, I suppose you will say that he sprays everything? Well, no. He sprays where it makes sense and brushes where it makes sense. For example, if the body of the house has been sprayed then they may roll the facias to avoid masking.

Spraying is not the secret, his systems are.

There are a few things that we need to think about when trying to improve our productivity on outside work and these are: your mindset, getting comfortable with spraying and planning your approach. We will talk about each one in turn.

Your mindset

This is the main one. If you decide that "normally" the outside would take a week and I tell you to complete it in a day, your mind immediately says "impossible" and you reject the idea.

This is normal, my mind works the same, but you will never do things better if you don't believe that you can. Let us just for a minute use another example of how our mind works.

How much could you earn an hour?

£15, maybe £20 or £30. This is what most decorators would say if they were asked about an hourly rate. What if I told you that I know a painter that earned £160 per hour on one particular job?

You may smile and nod but, in your head, you are thinking "bullshit".

Another question for you. How much is it possible to earn an hour? Not just a painter, anyone. For example, premiership footballers earn a lot. If you are on £200K a week then that is £5,000 an hour. Phew, that makes £160 seem like a low wage.

But in the world of business, even this is small change. Some venture capitalists can earn £1 million an hour. Yes, that is a million. Now suddenly the premiership footballers are roughing it.

Okay, back to reality.

We are limited if we work for ourselves and trade our time for money with a customer. However, we can get more efficient with our time, which will give us a greater return.

The point of all that is that sometimes our mind can put limits on us.

One more story and then I will get back to the main plot. For many years we all thought it was impossible to run a mile in less than 4 minutes. It was like a barrier that we as humans could not cross. A bit like the speed of light. A speed limit cast into the laws of physics.

Then in 1954 along came a fresh-faced Roger Bannister. He was only 25, and he decided that the 4-minute barrier could be broken, and he was determined to do it. His mindset said that it could be done, and he worked towards achieving it.

Once Roger Bannister showed the world it was possible, many athletes have now done it, thousands of runners have run the mile quicker than 4 minutes.

Why all of a sudden could everyone do it? Well, there is nothing like seeing it with your own eyes to change your mindset from one of not believing to one of believing.

If you do not believe you can do something then there is no way in the world that you will do it. You need to change your mindset first on our journey to becoming a well-paid decorator.

Back to thinking about how long painting the outside will take.

Whatever you decide will usually be how long it takes. So, if you think, "right I am going to do this in 4 days instead of 5", then you will plan the work to be done in 4 days.

It is a tough one because you do not want to rush too much and not make a good job, or even worse fall off your ladder because you are rushing. But you do want to push the boundaries on what you can do.

Getting comfortable with spraying

You do not have to spray outsides to be more efficient but let me tell you that it will help a lot. I can spray a window in a couple of minutes, and it would take a couple of hours (three coats) by brush. That is just one example of many that I could quote.

The main thing that puts people off spraying outsides (once they get their head around spraying, that is) is the dreaded overspray. We have all heard the story of a mate that sprayed an outside and covered a hundred cars with overspray. I saw one post on Facebook that someone sprayed an outside and there was paint on cars 2 miles away.

Two miles!

First, I would like to say that, yes, you can get overspray, of course you can. But many of the stories are from the old oil-based days when the mist would stay wet in the air for longer. With water-based paints after about 2 or 3 metres it is just a dry dust.

There is no way that a water-based paint could travel 2 miles and stay wet.

If you do the following five things, then you can cut overspray down to zero. The main things that cause overspray just happen to be the things that people new to spraying do. These are:

1. **Having your gun too far away from the surface.** You should be around a handspan away. If you let your gun drift too far from the surface, then you will get loads of overspray. A small change in distance has a large effect. So, 30 cm from the surface and you get very little, 60 cm and you will get loads.

   The problem is that a handspan feels too close, so most people pull away from the surface.

2. **Having the pressure too high.** This is another beginner error. Crank up the pressure, it does not matter they say, and the paint is literally bouncing off the walls. What the pressure should be set at depends on a few factors such as the type of paint and the thickness. Plus, the weather will have a bearing too. On a nice warm day, you will need less pressure. Typically, it will be about 2,000 psi but sometimes you can get it as low as 800 psi.

3. **Waving the gun around or arching.** For some reason, people think that it shows finesse if they wave their gun about like some mad sprayer dancer.

   The gun should be at right angles to the wall at all times. This is exceedingly difficult to do because your arm naturally fans. However, you need to try. Sometimes you think you are doing it right, but if you video yourself and watch it back then you might be surprised.

4. **A worn tip.** I cannot remember the number of times that I have had a sprayer proudly tell me that a tip has lasted years. Chris Berry uses a new tip on every outside. Yes, that is a new tip every day.

Why is that?

Because a worn tip creates lots of overspray and wastes paint. Not many sprayers know this and that is why we have so many overspray horror stories.

5. **Use a shield on the danger zones.** The danger zones are the corner of the building where it is easy to spray past the corner and onto cars. If you use the shield on the corner, then you will trap any overspray and prevent it from going any further.

As you can see from the picture below, the gun is behind the shield, so it is impossible for paint to get past the cardboard and onto the world of cars.

If you do all these things, then you will not get any overspray. However, the best thing to do is start on a job that is really safe on a day when there is no wind and build your confidence from there. Before you know it, you will wonder why you ever used a roller.

Planning your approach

This can feel strange to your average decorator but what I will ask you to do is write down every step you take when working on an outside. The best approach is to choose a job you are working on, and each day write down what you did.

Here is a simple example. I am assuming that we are just painting the facias and soffits and front door.

7:00 — Arrive at the job, unload the van and set up a base for the paint and ladders.

7:30 — Sheet up and have a brew.

7:55 — Rub down and prep all the facias, soffits and front door and casing.

10:00 — Brew.

10:15 — Carry on prepping.

12:30 — Lunch.

1:00 — Continue prepping.

4:45 — Tidy up and then finish for the day.

5:00 — Finish.

Today, I have prepped all the surfaces to be painted ready for undercoat tomorrow and gloss on the third day.

As you can see, I have included times and I have included any breaks and lunch, the goal is just to see how your day pans out. This is your **current system** even though you did not know that you had one.

Once you have done this for a real job you can then look at it and see how you can improve your current system. Your new system needs to be better. By better, I mean that you want to eliminate any steps that are not needed. You need to improve any steps that cannot be eliminated but can be done quicker.

You also want to eliminate any dead time.

These are things like fag breaks, giving the job a coat of looking at, checking out Facebook, texting your friends, checking out Facebook again and uploading pictures onto Twitter. As you can see, there is no end of time-wasting behaviours.

Now I know what you are thinking, you can do what you like, it is your business. Yes, you can do what you like, but we are trying to improve the bottom line here. It's up to you. Check Facebook or have money in the bank.

An improved system may look like this:

7:00 — Arrive at the job and leave things in the van. Sheet up.

7:05 — Start prepping using a power sander, fill as you go.

114

11:30 — Once prepped, mask off the wall areas under the soffit.

12:30 — Lunch

1:00 — Undercoat everything using a sprayer and a water-based exterior undercoat.

2:00 — Once dry (assuming great drying conditions), apply the first coat of gloss.

3:00 — Apply a final coat of gloss.

4:00 — Demask, tidy up and load up the van.

5:00 — Finish.

One of the many advantages of spraying is that the paint is smooth when applied so you do not need to abrade between coats. Also, the previous coat does not have to be completely dry before applying the next coat. With water-based coating on a nice warm dry day, the paint can easily be recoated in an hour.

This is just an example, but I think you get the idea. Have a little notebook and write down your current systems. This will give you something to work with for future improvements.

Managing the weather

Another big problem that we have in the UK is the weather. It is very unpredictable. You need to manage this. One advantage of doing the outside quicker is that you can do the job in a shorter weather window.

I have always found this to be a difficult one because our weather is so changeable. You do get good at reading the forecast and using the dry spells to maximum use. If it will be fine and dry for 3 days, then maybe work long days for those 3 days and get 3 outsides done. If you do it right this could give you enough money for the month.

Chapter 12
uPVC systems

"However beautiful the strategy, you should occasionally look at the results."

Winston Churchill

"Are you doing a course on spraying uPVC?"

This is a question we are repeatedly asked at the Academy. The reason for this is that it is an emerging market and there is quite a bit of misinformation out there.

While this is not the kind of work that I do, I know a lot about it and I know quite a few people that do it, so I will cover the basics here to show you what is involved.

Everyone is interested because there is good money in it. You could earn upward of £500 a day with the right work and the right systems.

This is a new market in the UK and there are loads of people jumping on the bandwagon. What I would say is that the market will eventually settle down. When this happens profits may not be so astronomical, but I think they will always be good.

If you are considering doing this in your business, you should specialise, let people know that this is what you do all the time (even if you do the odd decorating job between) and be really good at it. Do not cut corners for a quick buck, play the long game, do the prep right and use good products.

The following things are what you need to understand when painting uPVC. The products, the application method, the preparation and the pricing.

The products

There are several products on the market designed to paint uPVC and some that are just general "will stick to anything" products. I will not list all the products here but, instead, I will talk about three that I have experience of.

Kolorbond

This company have been spraying plastic windows for 25 years, so they have a real track record of doing it. They also have been selling the product that they developed to spray their work in Europe for the same amount of time.

They know a thing or two about it. If anyone is giving you advice about anything then the first thing to ask is what do they know and why. I feel that with their track record Kolorbond are worth listening to.

Kolorbond do three products:

1. Kolorbond Original. This is a quick-drying solvent-based paint that actually bonds to the plastic and becomes part of it. It is an amazing product that has to be sprayed. It is too thin and too quick-drying to apply by brush.

   It is the company's flagship paint and is tried and tested.

2. Then they make a 2-pack version called K2, for composite materials that is sometimes used for doors. The composite will not take the Kolorbond Original paint so that is why this product was developed.

3. A water-based version called Aquatek. This is for use on the inside of the window because the other two paints would be too dangerous to use due to their smell and low flashpoint. The water-based paint can be brushed or sprayed to the plastic window.

HMG paints

This is a well-established company based in Manchester. I have been on one of their training courses and I have to say that I was impressed with their culture and products. They make the paint used on the Millennium Falcon and that's good enough for me.

Seriously though, these guys are an industrial paint company that has been around an exceptionally long time. They have developed a range of uPVC paints.

Their solvent one is called PVC 17 and is designed to be sprayed. It is fast drying, is nice to use and is cheaper than Kolorbond.

PVC 17 is being replaced by a new product called "PVC pro" and is a one pack solvent based system. The product is designed "ready to use" without the need of additional thinners when spraying. There is going to be a water based version of the product too.

Here is a snippet from their marketing materials.

"As the leading UK independent coating manufacturer HMG Paints' PVC range is utilised across a range of plastic and difficult to coat surfaces, in a number of industries across the UK and Ireland. The latest addition to the range, PVC PRO, is a high-performance premium product designed for uPVC Windows and Doors. PVC Pro can completely transform UPVC window frames, doors and even conservatories."

Zinsser Allcoat

This is available in both solvent-based, and water-based paints. I have used both paints and I like them. Some people use this for uPVC windows, and it will work because it sticks to most surfaces.

You can apply it by brush and by spray, so it makes it a good all-rounder. Plus, it is widely available in most decorating suppliers.

However, I would not use it if I were going to be a uPVC painting specialist. This paint will not bond to the plastic-like coatings that are designed just for uPVC and I think it would be a compromise.

The application method

If you are going to paint someone's plastic windows, then you want them to look like they are brand new again and you're not going to get that applying the paint by brush.

The products designed to be used on uPVC windows have been designed to be sprayed anyway so you would struggle to apply them by brush.

The paints are thin, almost like water, and they dry quickly (seconds) so you would struggle with any other method. The good news is that products such as Kolorbond Original are amazingly easy to spray. It is the easiest part of the process.

You have two options of which system to use. You can use a conventional compressor and spray gun, and this will set you back a couple of hundred quid. Many people use this system and it works fine.

Kolorbond spray their windows in a spray booth and they use a conventional spray gun.

The disadvantage of this set up is that you get more overspray and it is cumbersome on-site. In a booth, you have extraction and you are not humping a big compressor around, but on-site it is different.

An alternative is an HVLP turbine. I have looked at these in an earlier chapter so maybe flick back and have a look. The advantage is that they are exceptionally light and portable. They are very expensive, though, and a decent system will set you back over a grand.

This is the system I would use. I have sprayed Kolorbond Original through an HVLP at college and it sprayed really well.

Because the paint dries so quickly, it is difficult to get runs. You can put the second coat on after 10 minutes and you are done.

The preparation and masking

This is where the work really is. You need to be sure that the plastic is properly prepared. Paint does not like to stick to plastic and if you do not prepare the surface properly then the paint could flake off. It is very unlikely that the paint would be at fault.

There is silicone around the window where it meets the brickwork. This is usually in a colour to match the window. So, if the window is mahogany uPVC, the mastic will be brown. This needs to be removed because the paint will not adhere to the silicone and also you want the silicone to match or it will look rubbish.

Next, you need to prepare the plastic. This is done in three steps:

1. Wash down with soap and water or a cleaning solution that the paint supplier provides.

2. Clean down with a solvent such as methylated spirit, acetone or white spirit. Some paint companies do a solvent cleaner that goes with their system.

3. Finally, abrade with a fine Scotch-Brite pad.

Then you need to mask. You will need to mask the glass off and also around the brickwork. The paint will stick to the rubber seals around the glass, however. I always think you can tell the windows have been redone if the seal is painted. A bit like having your car re-sprayed and painting the seal around the windscreen.

Mask the rubber and go onto the glass, too. Then go round the edge of the seal with a knife so that the line is perfect.

The rubber seal

Mask onto the glass and seal

uPVC systems

Use masking paper on the glass

Here is a picture showing a fully masked door (supplied by HMG paints)

The completed job (supplied by HMG paints)

The openers

Some windows will open and you need to spray them open, which gives you a problem. You will be exposing the inside of the house and you will be spraying all the window mechanism.

If you can remove the opener then this is the best approach, and then fill the hole left with masking paper or card.

Different window manufacturers use different opening systems, and some are easy to remove, and some are not. If you cannot remove the window, then you will have to do your best masking the open window.

Another thing to consider is the edge of the windows. You need not paint the edges as you would on a wooden window, you can see if the windows are mahogany where the colour ends.

You are now ready to spray. This will not take long, and the product should be dry to demask after a short time. Most uPVC jobs can be done in a morning if there are two of you.

The final job is to re-mastic around the edge of the window in a matching colour. If the windows are dark grey, then the mastic needs to match.

The pricing

I am not going to tell you how to price a uPVC job. If you do not know how to price then check out my pricing book, "Fast and Flawless Pricing".

What I will do though is give you some pointers. You need to price using a system. For example, so much per pane and extra for each opener.

This window has ten panes (five large and five small) and two openers.

To clarify, eight panes are not openers and two panes are. You will charge more for an opener because it is more time-consuming to do.

You may charge £60 per large panel and £30 for a small panel plus and extra £20 for an opener then that's:

£60 X 5 =	£300
£20 X 5 =	£100
£20 x 2 =	£40
Total =	**£440**

The point here is not to copy my prices but to make sure that you price this way. The price per panel will vary and it is easy to reverse engineer other companies' prices to see what they are charging per panel.

Just do not price on time.

A typical outside may have a price of £1,000 and you may complete it in a morning. If you told the customer that you charge £250 an hour then you will have a lot of hassle. Just tell them you have a formula to calculate the price and that the price is an industry-standard for reputable finishers.

Chapter 13

Recruitment systems

"Hire character. Train skill."

Peter Schutz

I know what you are thinking. Great quote, but who is Peter Schutz? Well, maybe Google him to find out.

I toyed with the idea of not writing this chapter and just leaving it at that, because it sums up what I am about to say.

What is important when taking someone on? How well they can paint? Do they wallpaper? How much do they want an hour?

None of these things are important. What is important then? I think the following:

- Honesty
- Integrity
- Respect
- Good timekeeping
- Clean and tidy
- Good people skills
- Willing to learn new skills

If someone is dishonest then that is it, you cannot train them not to be. Some qualities are essential in a decorator, they will be working in someone's house under the banner of YOUR company name. You want to be sure that they are enhancing your reputation not diminishing it.

Okay, it would be great if they could decorate too and ideally, we want the above list plus decorating skills, but we tend to focus on the decorating skills and forget the rest.

I think this is a mistake.

If you are going to build a company then it is important to recruit a great team. I have been involved with a company that has built its team from the ground up to a team of forty decorators in just over a year, so I have seen first-hand how this pans out.

If you have decided that you do not want to build a company and you will stay as a "one-man band" forever, then you do not need to read this chapter. I understand this thinking, it's pretty scary to start employing people and it introduces unpredictability into your business.

Most decorators will start on their own and then get really busy, the first step on the road to employing people is to think about taking on an apprentice.

I have a lot of experience with apprentices and employers taking them on because for over 20 years I used to teach them on their day release at college. Over the years I got to know all the employers.

A range of employers took on apprentices, from a decorator who took on his son as an apprentice to large firms who took on four apprentices a year.

From a company viewpoint, an apprentice is a great thing. They are cheap and typically will earn £100 a week in their first year so you can use the apprentice to do all the things that are expensive for you to do like preparation, moving paint around and cleaning up.

From the apprentices' viewpoint, they will get some real-life work experience and if they are with a good company, they will learn a range of skills that will set them up for life. If they go to college on a day release basis then they will get qualified too.

I speak to decorators, and they tell me horror stories about apprentices. They are useless, they cannot get out of bed, they are always on their mobile phone and they are lazy.

"Never again," they say.

But the trick with apprentices is to find a good one. If you get a good apprentice (and in fairness, there are good ones about) then you have a real asset on your hands. They will make you more cost-effective on your jobs and when they are qualified you have a decorator that you have trained and will do things your way.

I found that some firms were brilliant at choosing apprentices and some were rubbish. "I can't understand it," one company said to me, "we have had four apprentices now over the past 5 years, and they have all turned out to be rubbish."

Then I would have other companies that would take on three a year, and they would all be good. How can that be?

There is a skill in choosing a good apprentice. The main thing is that they are a good match for your company. You need to know what you are looking for in a person. It is different for different types of company.

Let us pick three aspects of people's personality and think about it. The three are intelligence, practical ability and how outgoing they are.

I think that the best apprentices are average intelligence, with a lot of practical ability and introvert (quite shy). I may be wrong with this and you would have to make your mind up what you are looking for but that has been my experience teaching thousands of apprentices.

If they are too clever, they will get bored quickly and will probably leave you once they are out of their time and set up on their own. If they are too thick, they will do stupid things and lack common sense. I told one apprentice to gloss the skirtings in his room, and he glossed the ceiling. You get the idea, imagine if he did that on a job.

Some people are just not practical. They have no ability. I had one student who was a lovely lad, interested in decorating and knew his stuff. He just couldn't do it. Even in his third year, he could not even cut in. He was just ham-fisted with everything he did. Some people are naturally practical, and some people are not.

Finally, the outgoing aspect of their personality. Now, this is a tough one because in some ways you want an outgoing apprentice as the quiet ones can be tough to work with. The outgoing ones chat to you a lot, chat to the customer and are generally "likeable", so they are great when you interview them.

However, they can become a nightmare because the social side of the job becomes more important than doing the actual job. These are the ones who, on a big job, just wander around all day chatting to the other trades and get nothing done.

The quiet ones, however, just get on with the job. They tend to be the ones that have thought about and want to be a decorator. I am an introvert and I love decorating because of it. I get to spend time on my own with no distraction and I enjoy transforming a room to a thing of beauty. I would rather work than chat.

This is what you want from your apprentice. They will come out of their shell as they get older, so they can deal with customers once they are out of their time.

When it comes to taking on apprentices, it can be a numbers game. Out of five apprentices, you may find one good one. The bigger firms over-recruit because they know this, they will just end up with one good one and get rid of the rest.

Sometimes that is difficult when you are a smaller company because you do not want to "give up" on an apprentice. You feel like you have failed. After 6 months you will know if they are any good or not and if they are no good then you are not doing them any favours by persevering, decorating is just not for them. They need to go away and find their thing while they are still young.

Taking on a qualified decorator is easier because they are already trained up and they have decided that this is their career and they have a track record.

Again, you need to have a set of traits that you are looking for. My three would be honest and reliable, a passion for decorating, and clean and tidy. But yours may be different.

They need to be reliable and honest. It's no good if you get a phone call every other Monday with a bullshit story about why they can't come in or why they will be late. You need to be able to trust them if they are working on a job on their own. "Will you be finished by Friday?", you ask. "Oh, I just need another day," they say. You need to be able to trust that they are not swinging the lead.

They need to like being a decorator. I speak to so many decorators that just moan about the job all the time. They hate preparation, they hate wallpapering, they won't spray because it is a waste of time. They hate rolling, too. I just wonder why they do not get another job. Postman, maybe.

When you work with decorators who really like the job, it is a different experience. They get on with it, they are happy in their work. They try to improve themselves and come with great ideas to do the job quicker.

Finally, you need to be a clean and tidy person to be a good decorator. Some people are just not tidy, and they don't even know it. You go onto their job and there is stuff everywhere, the paint is not in a neat pile, there are bits of sandpaper all over the place. Half drunken cans of pop and sandwich wrappers everywhere.

We are in the business of making a place look better, not worse. If you are a naturally tidy person then your work will reflect this. It will bug you if the line is not cut in right or if there is paint on the kitchen worktop. If you are an untidy person, you will just not notice.

How can you tell if they are a tidy person? I think you can tell. Just look at their appearance. Look in the back of their van. Look at a job they are working on.

To sum up you need to have a system of taking someone on. I would have a three-part system. This would be:

1. Know what type of person you are looking for as discussed above.

2. Have a good interview process.

3. Have a trial period of a week for experienced decorators and 6 months for apprentices.

Interviews are usually a phone call, you have a quick chat with the decorator and then say you can start Monday, this is no good. I would have a two-step process.

The first step is to interview them face to face. Ask them set questions. These could determine if they are honest and reliable, passionate, and tidy. You can also discuss pay and conditions and be sure that you are prepared to give what they are looking for and also that they are happy with your offer.

It amazes me when a new guy starts on the job and they ask me what they will be paid. "Did you not discuss that with the boss before you agreed to start?" I ask them. "No," they say.

What?

The second step of the interview is to get them to paint something. A door for example. Just say, "Okay, please will you paint that panel door?", and then watch them and see how they go about it.

Do they put a sheet down? How do they pour the paint? Do they prepare the door? How fast are they? Do they look like they can paint?

This simple 10-minute exercise will answer a million questions and sort out the bullshitters from the real deal. I would not tell them beforehand, either, so they can't do any research. If they are a decorator then they will know how to paint a door, if they are not then it will show.

I am amazed by how many companies take on a decorator without any kind of practical test.

As part of the interview, you want references. Not just one, either. You can ring up their old boss beforehand, so you know what they think before the interview starts.

A trial period is not unreasonable. A week is not long, you can pay them for that week, and it is long enough to see what they are like on a real job. I would put them with one of my most experienced and trusted decorators for a week and then ask them what they thought of the new person.

Will they fit in?

If the answer is yes, they were good, and they will fit into your company, then take them on.

Chapter 14

Pricing systems

"There are so many men who can figure costs, and so few who can measure values."

Anon

When I went back on the tools after working at college, I felt that the best way to have a profitable decorating business was to spray everything. This is what I did, I thought that was it.

As I spoke to other decorators that had sprayed, I found that there was something just as important that many of us were missing.

The skill of pricing.

I have written a whole book about this, so I will not repeat everything that is in there. If you are interested, then check it out. Search on Amazon for "Fast and Flawless Pricing". I am sure that you will enjoy it.

What I want to talk about here is having a system. Most decorators that I speak to price by estimating how long the job will take and multiplying that by a daily rate.

People are not particularly good at estimating how long a task will take. Not just decorating but any task. We always underestimate the time and we also tend to think the job will be done in ideal conditions with nothing going wrong.

Jobs are rarely like this. We hit snags and problems. The customer changes their mind, we run out of paint, the wallpaper takes twice as long to strip than we thought it would. The paint on the woodwork flakes off as we sand it.

You get the picture.

The longer the job will take the harder it is to estimate the time needed. So, for example, you may be able to guess well if the job is a couple of days. But if the job is 3 months then this is much harder, even impossible.

This is why it is important to have a proper system for your pricing. There are a few options out there. You can get pricing apps that allow you to enter different aspects of the job. The app has rates built-in so that it will spit out a price. If you want to enter your own rates into the app then you can do.

If you have your own rates, then you can set up an Excel spreadsheet and use that to quickly calculate your price from the areas and number of coats that you have inputted.

You can employ someone on a freelance basis to price jobs for you. This is not as expensive as you think and will give you consistent prices and save you the hassle of pricing. There is a company of quantity surveyors who specialise in estimating for decorating companies, and they are very good. The company is called P&D Online.

These guys will estimate for you on a job by job basis for a fair price, and they also have an app that you can download (for a fee) and use. I know a few decorators that use this company and they give exceptionally good feedback on their service.

I think it is good to understand properly how to price and get your head around it. If you are new to pricing, then it would definitely be useful to employ a freelance estimator. If you can find a local one or you know someone then that is even better. They can price for you, develop a set of bespoke rates just for your business and give you support when needed.

You will find that accurately pricing your jobs will mean you are more profitable, and this alone will pay for the estimator's time.

Chapter 15

Marketing systems

"Don't find customers for your products, find products for your customers."

Seth Godin

Before we start, we need to be clear about what marketing actually is. It is not selling. It is more than that, selling is just a small part of what marketing is.

I Googled for a definition because although I know what marketing is, I was struggling to put it into words. Here is what I found:

"It is the business process of identifying, anticipating and satisfying customers' needs and wants."

As decorators, we forget all about this. We have a skill, which is decorating, and we assume there is a need out there for it. Of course, there is a need otherwise the trade would not have developed in the first place. But this is just a general business.

Painting and decorating as a trade is a commodity in the minds of our customer, like sugar or tea. People are used to what a painter and decorator does, and they have a rather good idea what they charge too. They have friends who have employed a decorator and they discuss prices with them.

The problem is that commodities always bring the lowest price.

It is only by creating your own offer to the market that you can charge more for what you do. Part of doing this is to look at your potential customers and see what they want.

Maybe even try to anticipate their future needs.

Okay, I know, this is all a bit pie in the sky. Here is an example to make it clearer.

Painters have always painted outsides. At one time, windows were wood, and they needed painting every 4 or 5 years otherwise they rotted away. Most people do not like going up ladders because it is dangerous, so they got someone in. It was a real customer need and you did not need a marketing department to get the work.

Then something happened. Double glazing came on the scene. Plastic windows. Suddenly, everyone was getting their wooden windows taken out (even if they were still good in some cases) and getting plastic windows fitted.

The dream of no maintenance and a window that always looked good was a big selling point.

Double glazing salesman made a killing. A new need had emerged, and someone filled it. Not great for painters though, outside work diminished and there was less work to be had.

We all know that nothing looks good forever and plastic windows are no exception. Two things happened. The old-style white plastic started to look shabby and the newer style mahogany look windows started to look dated and old-fashioned.

Enter the decorator again.

You can spray uPVC and the windows look like new again. Now, this is not a new thing. A company called Kolorbond based in the Midlands, who make a uPVC paint, have been selling the paint in Europe for over 25 years, but it is only recently that we as a trade have cottoned on to it in this country.

The uPVC window sprayers (who often are painters) are making a fortune. Now the trend has become more obvious and more people are getting on board. In fairness, once the market matures, I still think there will be good money to be made in this niche if you provide a great service.

Imagine, though, if you saw this coming 10 years ago. That is what I mean by "anticipating customer needs" and this is part of marketing and is what market research is all about.

In the old days, most decorators would do two things to get work. They would use word of mouth and they might place an advert in the local newspaper. These two approaches can still work but with a little change.

First, "word of mouth" is a powerful method of selling your services. It is powerful because your current customer is recommending you to their friend. Their friend trusts them, so they will take the recommendation on board and give you a chance to quote for work.

A good idea to formalise this into a system is to have a reward for the person who recommends you. You can tell your current customer about the reward system or you can keep quiet and just reward them when they recommend you.

If you get a call from a new customer, you will ask them where they got your number from. This is always worth doing because I am always wary of unsolicited calls.

If the call is from someone who has got your number from a current customer then you can post a "thank you" note with a £10 voucher to your current customer. This will be a nice surprise for them, and you can bet your bottom dollar they will recommend you again.

It's always good to reward the behaviour that you want from your customers. You may make the reward in proportion to the size of the job. If the job turns out to be a well-paying £20,000 job, then a £10 voucher might be a bit lame. Maybe send them a nice hamper.

Use your imagination a little. If you know they like whisky, get them their favourite one. Nice personal touches like this go along way with people.

Using your local paper can still work. Many older people still like to read the local paper and if you're in there it could lead to work.

In my area, we have a company that produces a booklet which profiles local businesses. It is well produced, and businesses pay to be in there. The booklet targets more affluent areas and customers are encouraged to keep the booklet and use it to look up local tradesman. This can work very well.

Social media

We cannot talk about marketing these days without mentioning social media. It's the future, we are told. There are various platforms out there and the landscape is constantly changing, and some platforms come into fashion and others disappear.

Does anyone remember Myspace? Well, it is still going but no one talks about it. What about Tik Tok? Well, it is emerging, and people are using it more and more.

More current mainstream social media platforms are Facebook, Twitter, YouTube, Instagram and LinkedIn.

If you are going to have a presence on these platforms, you need to portray a consistent image of your company. Same logo and wording used across all the platforms so that people recognise you easily.

I think social media is a double-edged sword. On the one hand, it does, in theory, give you access to many potential customers and it allows you to showcase your work to a large audience. Our trade is a very visual one and it does lend itself to photos and videos on social media.

You can pay for advertising on Facebook and it is targeted to your selected customer base. This can work really well.

Another good aspect of social media is the opportunity to network with fellow decorators and swap good practice and get ideas for new things to try. Facebook forums are good for this. If you search for forums, you will see that there are a few out there. For spraying information and good quality advice, it's worth checking out "Spraying Makes Sense". These are free to join and if you're shy then you can just sit in the background and observe.

Is there a dark side to social media?

There are a couple of negative aspects to social media. The first one is that it can be so time-consuming, to the stage where it can take over your life. You innocently join a Facebook group that you are interested in and before you know it you are on Facebook for most of the day and evening.

It creeps up on you so that you do not realise that you are doing it. While you are in phone land, your perception of time changes too. What seems like 10 seconds is really 20 minutes. I see people on-site stood frozen, transfixed by their phone, reading some post about what someone had for breakfast.

I do not take a smartphone to work, I have a Nokia 3310 which only texts and takes calls. Some people find it amusing. My students once offered to buy me a "proper phone" like I could not afford one. Some people (for some reason) get angry about it and it pisses them off that I don't have a smartphone.

These people seem to think that you cannot live in the modern world without a little computer stuck to your face all day and all night. I remember a time when even mobile phones did not exist let alone smartphones.

Another aspect of social media is that it can start to control what you think. This happens without you realising. Facebook know, they have people that pull the strings (for a fee) and I suspect that it is more powerful than we actually imagine.

I was at a Facebook marketing training session. It was at a nice hotel and the large table had a collection of local business sat around the table. The discussion got around to how Facebook can mess with your mind and cause depression and the consensus was that as long as businesses were making money then what does it matter. I found it very difficult to stay quiet at that point, but I think "No wonder the world is the way it is" slipped out.

No one heard me though, I think. Or maybe they just ignored me. That is more likely, I think.

People can be a little nasty, too. They say things that they would never say to your face and there can be a polarisation of opinion, which is never a good thing. For example, in the world of decorating, "should we use masking tape or not?" Like there are only two camps. The maskers and the guys that cut in with a brush. You are in one camp or the other, you had better decide.

But you can cut in sometimes and you can mask sometimes. You can be in both camps or neither. But that option is never discussed. This tends to be the case with all "issues" that are discussed. There are rarely just two options. In the real world, there could be fifty options.

To sum up, social media has a place in your marketing effort. You need to have your act together; it is worth reading a book on it or doing a course.

Don't let it suck your brains out though and take time away that you should be spending with your kids while you argue with some stranger about water-based paint and if it's better than oil.

Chapter 16

Training systems

"It's all to do with the training: you can do a lot if you're properly trained"

Queen Elizabeth II

I feel that I know a lot about training because I have been a lecturer in painting and decorating for over 20 years, and I am also the director of a private training organisation called "PaintTech Training Academy".

I believe that training is the key to a great business, not because I want you to come and spend money at one of our training centres, either. I have always believed this, and it is the very reason I do what I do now.

I feel that we have been let down as a trade by the education system. I am having a go at myself when I say this, so if you teach painting and decorating in a college, please do not take offence. If you went back in time 10 years and asked me what I knew about painting and decorating I would say that I knew everything. If you asked me if I felt that the courses that my college offered were current and useful to get people into the trade, then again, I would say that I thought they did.

Then I left college and went back on the tools. I had a shock, I can tell you. The world had changed, and no one told me. There are so many things that are different now. There are more products available, there a wide range of good quality water-based paints too.

Customers are fussier and they want a top-notch job doing, no more just bashing a job out and getting paid. There were good changes, too. Decorators are in demand, there seems to be a shortage of good quality tradespeople.

I decided to get into spraying and when I spoke to people, I found there is a distinct lack of knowledge out there.

Where did I go wrong at college then?

It's difficult to understand if you have never worked in a college but what happens is this. At first, you are current in your trade, you have only just got off the tools, so you have lots of stories to tell and your knowledge is current.

Then you get sucked into college land. You're into meetings, teacher training, writing lesson plans and schemes of work.

It does not take long and really you are no longer a decorator; you are a teacher. You used to be a decorator.

Years go by and you get out of touch. You do not realise it, though, there is no real way of knowing. The syllabus given to you is written by people who are out of touch too, so nothing changes.

Another problem that you have is something called "space utilisation" that colleges (or government, I suppose) have. This is the idea that a room must always be in use all of the time. So, if you have a painting and decorating workshop then it has to be full all the time or they will take it off you. That is easy for the joiners because they have benches and the student puts their job away and another student can use the bench.

But for decorators, we decorate the room and it is in use.

If you did a "space utilisation" on your house, then what would happen? Well, you would get rid of the bathroom for a start because you're only in that a few hours a week. The kitchen too, probably. You see, it is a stupid system and typical of the college approach.

This meant that we were always short of surfaces. In the end, you have to design exercises that involve the student painting a small area really slowly so you can keep them busy all day. The exact opposite of what you need to be teaching them.

I used to design elaborate cutting in exercises that tied a student to a 2.5 metre by 1.5 metre panel all day. The students loved doing them, but it was not teaching them anything other than setting out, measuring, and cutting in.

Another problem at college is that the college is funded per student so there is an incentive to get as many students on courses as possible.

I actually agree that we should be funded per student, what I don't agree with is the blind expansion of the college to serve a few people at the top.

Students should be selected on merit and you should only train ones that want to learn. Another problem with college is that if they lose a student then they lose some of the funding. Again, there is massive pressure on the tutors to keep hold of students no matter how big of a dick head they are.

Another problem is keeping up to date. The latest kit that all the trades use these days costs money. In fairness, our last head of department was very future-facing and secured money to buy a load of Festool equipment and a new sprayer for the decorators. This is not the case in all colleges, though.

These factors and more are the reason I said goodbye to my college job and got back in the real world. I am glad that I did too because I have learnt so much over the past 5 years.

Speaking to many decorators all over the country I have come to realise that there is a shortage of good quality training. I think that the government want the training of trades to move away from colleges and into the private sector.

Once I was back on the tools and started spraying, I had many requests from decorators to go on do some on-site training. At the same time, I got speaking to a decorator in London called Ian Crump. He wanted to set up an academy to train the industry how to spray safely and correctly.

PaintTech Training Academy was born. Check us out at www.painttechtrainingacademy.co.uk

We did this and it has been remarkably successful. We have teamed up with City and Guilds and become an assured centre. This means that City and Guilds check that we deliver quality training programmes and in return, the students get a City and Guilds certificate.

We also teamed up with the CITB (The Construction Industry Training Board) and became an ATO (Approved Training Organisation). This meant that we could offer some of the CITB standardised short courses designed for the decorating industry and the decorators could get funding for them.

Many decorators do not realise that if they are CITB registered they can claim up to £5,000 per year for training their workforce. This can be claimed every year, too, so it is a real bonus if you are trying to build a well-trained workforce.

One reason many decorators do not apply for the money is that the paperwork can be a bit bewildering. Lucky for us, we are used to working with the CITB and filling in the paperwork so if anyone needs any advice we can help with the form.

If you are not CITB registered but you use some sub-contractors, then you can register and be eligible for grants. Many decorators worry about the levy and think that they will get hit with a massive bill. However, the levy only starts if your wage bill is above £80,000 per year. Then the levy is only small.

Another area we want to be involved with is apprenticeship training. This is something that I have a lot of experience of and I feel that we could do an outstanding job for the industry.

What would I do differently than the college?

Well, first, I would limit the numbers on the course to around ten students. That way, we could give lots of attention to each student.

I would also be very selective of the students that I allowed on the course. There would be an interview and only the students who convinced us they were serious about becoming a decorator would be allowed on the course. Places would be very limited so the course would be oversubscribed.

Companies would have to commit to training, too, and only companies who demonstrated that they would give the apprentice a good experience at work would be allowed to send apprentices to us.

Getting on the course would not mean that you cannot be thrown off it. Students that became disruptive would be removed from the programme. In my experience, if students see this then the rest buckle down and get on with it.

The workshop space would be big enough to provide a good working area for each student and the students would get to use the latest technology. We have some of the best sprayers available and a range of sprayers from different manufacturers.

We also have a first-class wallpapering workshop and some of the industry leaders teaching wallpapering.

Just a final note on colleges, I do feel that the staff that work in painting and decorating departments do their best for the students and I have worked with some real top-notch decorators over the years in college. I know staff from most the colleges in my area, and they are committed to doing their best for their student. I am by no means having a go at the staff.

However, I know that they will agree that it has become a thankless job and many are leaving for a well-paid hassle-free life on the tools.

If you are a lecturer reading this and you have some tales to tell from the college where you work then email me on pete@fastandflawless.co.uk and let me know, it would be nice to hear from you.

If you decided that putting a load of systems in place was just too much hassle for you but you decided that you would make just one change to your business then training is the one to go for.

Continually training yourself and your employees keeps you current and it also keeps you motivated. We all like to learn new things especially if they make our life easier. If you and your team do two or three courses a year, then it will make your company extremely competitive.

Chapter 17

Why are we in business?

"It's a matter of having principles. It's easy to have principles when you're rich. The important thing is to have principles when you're poor."

Ray Kroc

Why?

This is a question we ask all the time when we are 5 years old and we are bugging our parents to bits, but it is something we stop asking as we get older.

However, it is the most important question of all.

Why are we in business in the first place? It is different for each person. Some are in business to make more money than they did when they worked for a company and that is it. Nothing else matters. It is just the money. That is your why.

Some are in business because they want more freedom and control over their lives. "Freedom" is a common answer to this question. However, if you are in business you will know that the freedom thing is a bit of a myth and all that it means is that you are free to work 7 days a week.

Before you do anything to change your business, you need to know the answer to this basic question. For example, if you are in business just to make money or profit then working 7 days a week is not a problem. But if you are in business so that you may have a month in Tenerife then this needs a different approach.

What you want will also change, so what you want as a 20-year-old will differ from what you want as a 50-year-old.

We need to keep asking ourselves this question.

If you have just set up on your own, then the first thing you need to think about is what kind of business do you want. There are a few options, but to keep it simple I will look at four.

1. Just you.
   This means just keeping a few customers and serving them yourself. You can do a good job and really look after them. This is the simplest business and in fairness, you will never make a fortune, but you will not get loads of hassle either.

   It is easy to control.

I speak to many decorators who employ ten decorators, and they say that some days they wish they had never bothered and just stayed as a one-man band.

2. A small team.
    This would be around four or five decorators. I think if you can get some good decorators that work together well then this can be the ultimate business. It allows you to tackle bigger jobs and make bigger money by without the hassle that a bigger company can bring.

    It also means you can stay on the tools if you like but, because you have a team, you can take some time out (a month for example) and things still get done.

3. A large company.
    A multimillion-pound decorating business can be many small businessmen's dream. This is possible, and I know quite a few examples of this. You're going to have to do a lot of growing as a person to get to this level, though, and you will not be on the tools.

    You will need a few decorators, too, maybe fifty to a hundred although I think it could be done with twenty-five productive workers and the right kind of work.

4. A lifestyle business.
   This is all the rage these days especially in the world of internet business. You have a small business that you put a little time into, let us say 10 to 20 hours a week. This pays well enough to fund whatever you want to do with your life.

   Probably easier to do with an internet business but I think it could be done with a decorating business if it were structured correctly.

   If this sounds interesting, then check out "The 4-hour Workweek" by Tim Ferriss. It is a great read.

How are you going to structure your time?

If you worked for a decorating company, how many hours a week do we think we would work?

I am guessing 40 hours.

How many hours a day? I am guessing 8 hours.

What about, how many weeks a year? Again, I am guessing 48 weeks.

How do I know this? Well, it is standard, that is how. Where did this standard working structure come from, I wonder?

Back in the 1890s, workers in factories worked as much as 100 hours a week! 10 to 16 hours a day.

In 1926, Henry Ford introduced a new working pattern in his factories. An 8-hour day and a 40-hour week with no change in wages.

Why did he do this?

Well, believe it or not, he knew that if people had time off then they would use that time to spend money. They might even go out and buy a car and go out to the seaside at the weekend.

So now you know that the basic work structure that we all follow has been around for almost 100 years and no one has thought to change it.

Personally, I think it is time for a change.

Some Scandinavian countries now work 4 days per week because it makes their workers more productive. Yes, that's right! They make as much stuff in 4 days as they previously did in 5.

The upside of this is that they get a 3-day weekend. How good is that!

Hang on a minute, though, how can that be? You can't produce the same in 4 days as you do in 5, can you? Well, yes, you can. Human nature tends to make the work fit the time allotted to it.

If you decide that an outside will take you a week it will, if you decide it will take you 4 days then again it will. This is the main reason for this book. If we get our act together, put some systems in place and get more productive, then we can have 3 days off at the weekend and make more money too.

You will be more rested and keener to get going again on Monday. There is less chance of burn out or being pissed off with working.

When the coronavirus lockdown occurred, the site that I was working on was closed. I had a few weeks at home with basically nothing to do. I was climbing the walls and when we were allowed back on-site, I was thankful, I can tell you. So, you need something to be working on, but you just need to try and not do too much.

It can be a difficult balance when you are self-employed, but it is something that you need to think about.

So, before you put loads of systems in place, decide on what you want first and then get to work.

Chapter 18

Can systems be a problem?

"Dangers lurk in all systems. Systems incorporate the unexamined beliefs of their creators. Adopt a system, accept its beliefs, and you help strengthen the resistance to change"

Frank Herbert

We are getting to the end of the book on systems and how they are good for your business. What I want to look at before we finish are the potential problems that systems can cause.

Let take a little trip down "systems lane" first.

You have been running your decorating business for a while now. You are still a one-man band and you mostly make it up as you go along. You do not have a marketing system and you price by guessing and you have no real marketing strategy, you mainly get your work from word of mouth and referrals.

Nothing wrong with any of that but you feel you have reached a plateau and you are not progressing any further. You make a reasonable income from your business, but you are not breaking any records. Your mate who works in a factory makes more money than you do.

You decide that you want to take your business up a level. So, you look at the current market for decorators and decide to specialise in wallpapering, with a focus on murals for corporate clients. You can continue to do your bread and butter work until the new venture gets going.

The reason for doing this is that you feel that if you offer wallpapering then you can charge more for your services because this is something that customers cannot do themselves and also it is something that not all decorators offer so it makes you different.

Even more specifically you want to offer a mural installation service to mainly corporate customers. You feel that the focus on murals is on trend and it is emerging as a market especially for larger companies who want to give a unique feel to their customer-facing areas.

You start to put a few systems in place.

First, you put together a marketing system. You identify your ideal customers, and they are:

Hotels

Apartment buildings

Restaurants

Can systems be a problem?

Fast food outlets

Retail outlets

Pubs

Expensive domestic properties

Nursery schools

Universities

This is quite a list! The next thing you do is contact local businesses that fall into the above categories. You are looking for the person who holds the purse strings for the redecoration work.

Some of the list, for example hotels, will be national companies and their account holders will be more difficult to get hold of. However, there will be local hotels that are privately owned.

The object of this marketing exercise is not to sell your product but to get feedback on what they think about the offer.

You could offer a mural for the feature wall in their reception, this would be a bespoke image that means something to them. The cost would include the mural and the installation.

The object of the exercise is that you want to understand if they would:

1. Want something like this.
2. Be prepared to pay a premium for it.

That's it.

You are just looking for feedback from real customers that you can use for your new marketing system. At this stage, you could offer attractive discounts if the hotel lets you use them in your marketing. You would both benefit; they get an amazing mural in their reception for a great price and you get a real-life case study and an endorsement from the customer.

Once you get through your list on a local level, you can take stock. I will assume that this process went well, and you have five customers who want to be used as case studies.

You carry out the work and take lots of photographs and video. You also get quotes from the customer saying how great you were and how it has transformed their business.

Now that you have done some market research you know there is a market for what you do, and you also know what they are prepared to pay. Now you can put your marketing system together.

Before you start your marketing efforts, you need to make a few decisions. You need a good name for your business. You are currently called Dave Smith Painter and Decorator, which in fairness was great for doing the local decorating work but would give the wrong impression for the corporate clients you intend to reach.

You decide to call your business "Custom Mural Installers". This tells the customer what you do and creates your own market for pricing and standards. I discuss this in much more depth in my pricing book, so I will not go over it again here.

You then get a logo designed. You do some research and find you can design one yourself. However, you think this may be a bit amateur, so you pay someone to design your logo.

The next thing you do is buy some website names or URLs that fit with your custom mural niche. You may not use all of these straight away, but it takes them off the market for your competitors and also gives you future options.

You find that www.custommuralinstaller.co.uk is available so you buy this. URLs are cheap to buy so this is not a big commitment.

You set up your website with your new company name, your logo, a description of what you can offer companies and some case studies with photos and feedback from the five customers you have done work for.

This is a great start, you have established your name, a logo, a website which will be central to your marketing efforts and some good basic content.

You have come a long way.

The next step is to drive traffic to your website. This will do two things. It will improve your site's visibility to Google and you will have more chance appearing in a search and also it will get people looking at what you have to offer.

You decide that you will market through these channels:

- Trade publications
- Trade bloggers
- Facebook
- YouTube
- LinkedIn

Using the internet for marketing is a two-edged sword as we have previously discussed. On the one hand, it can be cheap and targeted and have a good reach, globally if you want that. But you can spend all your time online and not get any work done.

Personally, I would allocate a set amount of time to it and try to do it in "work" time and not let it eat into your evenings too much.

Using social media to promote your business is a book in itself or a course. It is worth spending some time getting your head around it.

Let us assume your marketing system consists of:

1. Writing a blog post once a week and promoting it to your followers on social media.

2. Posting a video on YouTube once a week so people can see how amazing your murals look.

3. Showcasing some of your work on LinkedIn and interacting with companies that you follow that may use your services.

4. Making cold calls once a week to your target customer.

I will leave it at that.

The main thing is that you have a system and you carry that system out. You may find you start to have more work than you can handle. At this stage, you would need to build a team and look at your recruitment system.

You get the idea. I won't take this any further, I just wanted to talk about how the systems journey would start and how it would feel.

Let us imagine that we are 5 years into our new business. Things are going well: you have a team of installers and you are busy. The bank balance is healthy, and you have even taken someone on part-time to manage the marketing.

They are following your carefully designed system and to be honest you just leave them to it and don't even think about it. The same with all the systems you have going. You have found that you enjoy more spare time these days and you would rather be out on the golf course than checking up on your marketing person.

That right there is the problem.

The good thing about systems is that you can automate your business but things change over time.

You would notice the change because it's your business and it affects you financially. However, a person that you employ will not. They will blindly follow your system to their grave.

For example, the blog may be poorly written, and it is not driving any traffic to your website and none of those visitors are converting to paying customers. The marketing person may like using Facebook but in 5 years' time, Facebook may have fallen out of favour and no longer brings in any work.

The cold calling may bring in most of the work, but the marketing person doesn't like doing that (they feel it is beneath them) so they only do one call a week.

This can be the problem with systems, too, in that people who work for you will bend the system to meet their own likes and dislikes and not do things that maximise what you are looking for: more customers and more profit.

You need to review your systems and you need to know what your system is meant to be achieving. So, for example, if the reason for writing a blog post is to drive 1,000 people to the website then measure this and check it. If the numbers fall over time, then it can be looked at.

You may cold-call a hundred customers and this brings in five new jobs but, over time, this number falls (or increases) then you need to look at this. Do we do more of this if the numbers are better, or do we change our approach?

You may find that Facebook is viewed as old-fashioned (in the future) and a new kid is on the block, so you switch your efforts across to that. There are only so many hours in the day, so you don't want to waste your efforts.

You get the idea.

This is just one system. Your pricing system will need to be reviewed as well. As market conditions change, you may be able to put your prices up due to high demand or you may have to drop your prices due to recession.

You cannot just blindly follow the system just because it has worked in the past.

This happens to large companies. They start small and they do well because the small team that started the company are developing things and they know what they are doing.

In time, the company gets big and is run by employees, the original owners are long gone and have sold out to shareholders. Then times change, the company does not and before they know it, it's too late and they are gone.

Do not make that mistake. If a system is not working, then get rid of it. If it is taking too much time, then rethink it. It is easy to get sucked into it all and get carried away.

Times change, we need to be aware of the changes and we need to change with them when needed.

I speak to many decorators out there who have been in business for many years, they are happy doing what they do, they brush and roll out jobs to a low standard, they avoid wallpapering and they have a scruffy van.

Times have changed and they have not noticed. In fairness, they may not know, and ignorance is bliss, or they know but they are 5 years from retirement, so they will not change now.

To sum up, you need to keep up with what is going on in the industry, you need to put systems in place so that your business runs well and you need to review your systems so that they keep working well. Only put systems in place that you need, do not get carried away.

You also need to make sure that your employees are following the systems that you have set up correctly and are not freestyling too much. This is one reason that at college we had "mystery shoppers" who would come to enrolment and make sure that we were all following the correct process.

Chapter 19

The last one

"Don't water your weeds."

Harvey Mackay

Well, here we are at the end of the book.

I hope you have enjoyed it so far. Before I go, I want to sum up the vision that I have for our industry.

Many years ago, in my area, there were quite a few what I call "proper decorating companies", by that I mean they had the following characteristics.

They were established with a well-known company name and branding. They employed decorators (not sub-contractors) and they paid their decorators holiday pay and sick pay.

They also trained their decorators, they sent them to college to do an apprenticeship, and they made sure that they had a wide range of experience when they were at work.

They took pride in their work that they turned out as a company and their reputation was very important to them. They were proud of what they did, and they strived to get better at it.

These days the landscape has changed.

We have one-man bands or small companies with teams of two or three decorators. There is an army of sub-contractors that work for the larger firms and builders.

Then there are the larger firms. These are national multimillion-pound companies. Their main focus is to make money for their shareholders. There are exceptions to this large company approach but in my experience, they don't treat their decorators well, the work is not very varied, and they are burdened with a raft of petty big company rules.

These big companies would have their apprentice painting fences for 3 years. They are good at the talk, though; they will go on record to say how they care about the industry and how they train apprentices and then their behaviour shows how they only care about money.

I will not name any names because I think that all big companies are the same, but I worked as a sub-contractor for a big company for a short while.

Before they would give me any work, I had to do some training at my expense. But, to satisfy their policies, I had to

The last one

wait a month for my money, and they deducted a processing fee from me.

While I worked for them, even though they showed no commitment to me, I was expected to drop everything at their whim and be there at a moment's notice. Their philosophy was: "If we say jump then you say how high."

During the time I worked for them, I felt that I was a nobody and I meant nothing to them, step out of line and I was gone. A massive difference from my old family firm where I served my time.

I did not stay with the company long and I sent a nice letter to my foreman telling him thanks, but no thanks. I think he was surprised that I did not want all the work they were offering.

I know someone who works for the same firm, and he was asking me if I knew any decorators because his firm was taking on. He has been asked by his boss to ask around because they were struggling to get people.

I wonder why.

These big companies have done a lot of damage to the decorating industry. We need to turn the tide and build "proper decorating" companies again. Smaller and more nimble companies of maybe five to ten well-trained decorators with a passion for the trade.

A company with proper systems in place and with decent ongoing training.

The world is changing, and the tide has turned against bigger companies. You no longer need to be big to get access to decent systems any more. The internet has created a world where we can all be a media outlet and we can all be a publisher. All you need is a computer, and these are cheap and powerful these days. Your mobile phone has more computing power than NASA had when they put a man on the moon.

Large companies need to be scared because the new breed of smaller well-trained decorating companies will run rings around them.

A large company has a social media output that is embarrassing. A small company has its finger on the pulse. A large company will struggle to implement new systems, a small company will do it the next day.

The employees at a large company feel that they don't matter to the company. The employees of a small company are the company's number one asset and they feel like they are.

Decorators ask me if I think that every decorator will be spraying in the next 5 years.

The short answer is "no" but what I think will happen is that the trade will splinter into two.

One group of decorators will be "old school", they will carry on brushing and rolling, guessing their prices, and making the job up as they go along.

A new breed of decorators will emerge, and they will embrace new technologies and they will have proper pricing systems. They will plan their jobs so that they are two, three or four times more productive. They will recruit good, well-motivated staff, and they will train them so they are even better.

What will be the difference between the two groups? The new breed of decorator will make much more money and get greater job satisfaction and security. They will also have more work than they can ever handle and actually be in a position to choose who they work for.

Which group do you want to be in?

Other books by the author

Fast and Flawless

A guide to airless spraying

This is a chatty guide to airless spraying for decorators, decorating students and anyone interested in spraying with an airless system. The book covers all aspects of the airless sprayer including the components of the system, the different systems that are out there to buy and setting up the system.

The book covers topics such as types of sprayers, essential equipment, using the equipment, masking, PPE and masks, a bit about paint, what to do when it all goes wrong, spraying in the real world and common paint defects.

Fast and Flawless Pricing

A guide to pricing and business for decorators

Are you a decorator that struggles with pricing?

Have you just set up in business and are looking for some pointers?

Are you an established business looking for some inspiration on how to move forward?

This chatty guide on pricing and business will gently guide you through the process of pricing a decorating job. It looks at the pitfalls of getting your pricing wrong and the advantages of having a good pricing system.

The book has been written by someone who has both been a decorator and taught decorating in a local college for most of his life.

Tales from the building site

Lessons learned when working on a big site

The author has spent many years working in the building trade as a decorator. During all those years he has seen things that have made him laugh and things that have made him tear his hair out.

There have also been many occasions that have made him proud to be part of it all. Here is a book for all you people in the trade and also for everyone else who wonders what goes on behind those big high hoardings that clearly state the public are not allowed in.

It is a warts and all look behind the curtain from the perspective of a decorator. Be prepared to be shocked, to laugh and to shake your head in disbelief.

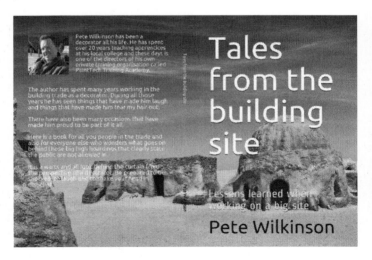

Boat Life

The trials and tribulations of living aboard

Nothing to do with construction or decorating. I love boats, I have one and I have lived aboard myself, so this is an insight into the lifestyle.

This is a book for Boaters, written by a Boater. Pete Wilkinson has spent his whole life around boats and has owned a couple too.

The book looks at all aspects of boating including, what is the best boat to buy, where to look when buying a boat and do you build one or do you buy one?

The following questions are answered: Which is the best type of boat – wood, fibreglass or steel? Do you borrow money or save to get your boat? Do you move around or stay put? What is the essential kit needed? How do you keep her shipshape?

The book also looks at living aboard and gives an insight into the liveaboard life. The book also puts boating into the context of modern life and discusses the advantages of living onboard.

Finally, if you have wondered what goes on behind the curtain of a Boater's life then this book will show you.

Check out the website

If you are interested in being kept up to date with future books, or you just fancy the odd freebie, then subscribe on my website.

www.fastandflawless.co.uk

About the author

Pete Wilkinson has been a decorator all of his life. In his younger years he worked for a medium sized decorating company doing a wide range of work.

Then at the age of 27 he got a job teaching Painting and Decorating at a local college. These days he runs his own training company called PaintTech Training Academy.

When he is not working, he likes to spend time relaxing on his boat with his wife Tracey.

Printed in Great Britain
by Amazon